Student Study Guide

for

Press, Siever, Grotzinger, and Jordan's
UNDERSTANDING EARTH
Fourth Edition

Peter L. Kresan
University of Arizona, Tucson

Reed Mencke, Ph.D.
University of Arizona, Tucson (retired)

W. H. Freeman and Company
New York

ISBN: 0-7167-5776-1

© 2004 by W. H. Freeman and Company.

All rights reserved.

Printed in the United States of America

First printing 2003

W. H. Freeman and Company
41 Madison Avenue
New York, NY 10010
Houndmills, Basingstoke RG21 6XS, England

www.whfreeman.com

Contents

Part I: How to Study Geology
   1 Brief Preview of the Study Guide   1
   2 Meet the Authors   4
   3 How to Be Successful in Geology   9

Part II: Chapter-by-Chapter Study Resources
   1 Building a Planet   19
   2 Plate Tectonics: The Unifying Theory   26
   3 Minerals: Building Blocks of Rocks   35
   4 Rocks: Records of Geologic Processes   45
   5 Igneous Rocks: Solids from Melts   52
   6 Volcanism   66
   7 Weathering and Erosion   76
   8 Sediments and Sedimentary Rocks   86
   9 Metamorphic Rocks   96
  10 The Rock Record and the Geologic Time Scale   106
  11 Folds, Faults, and Other Records of Rock Deformation   118
  12 Mass Wasting   127
  13 The Hydrologic Cycle and Groundwater   136
  14 Streams: Transport to the Oceans   147
  15 Winds and Deserts   156
  16 Glaciers: The Work of Ice   163
  17 Earth Beneath the Oceans   171
  18 Landscapes: Tectonic and Climate Interactions   180
  19 Earthquakes   188
  20 Evolution of the Continents   198
  21 Exploring Earth's Interior   209
  22 Energy and Mineral Resources from the Earth   220
  23 Earth's Environment, Global Change, and Human Impacts   229

Appendix A: Eight-Day Study Plan   239

Appendix B: Final Exam Prep   241

Answers to Practice Exercises and Review Questions   243

CHAPTER 1

Brief Preview of the Study Guide for *Understanding Earth*

We know from personal experience that studying geology can be very rewarding. Geology allows you to look at the Earth around you and understand how it came to look the way it does. It offers scientifically sound explanations for geologic disasters, such as earthquakes and volcanoes. As author Tom Jordan tells his students, "I'm going to give you a new set of eyes with which to see the world." We also know that mastering the ideas of geology can be challenging. As teachers, our main goal is to help you meet that challenge. We think you will find the organization of this guide both practical and helpful. Success in geology revolves around lecture, as do the study aids in this guide. Specific geology study strategies, including how to take an excellent set of notes and how to prepare for exams, are woven throughout this guide. Study aids are presented step by step (see flowchart on page 3) to enable you to know what to do before lecture, during lecture, after lecture, and during exam preparation. A final appendix—**How to Study Geology**—brings all these aids together into a short course on effective study strategies. Here's how to use these materials.

Before Lecture: Preview

The key to getting a thorough set of notes is to come to class with an overview of what will be covered already in mind. That way you will know what geological processes will be explained and what key questions the lecture will answer. In lecture, you can *actively* listen for answers to those questions. Listening actively with specific questions in mind is guaranteed to result in understanding more of what is said and getting more of the key points into your notes.

The method you use to gain such an overview is called **Chapter Preview**. To make previewing easier for you we begin all chapters of the Study Guide with three or four key questions each chapter (and its corresponding lecture) will cover. Brief answers designed to help you start thinking about the material are supplied as well. Working with these ahead of time will bring you to lecture ready to listen and take a good set of notes.

During Lecture: Take an Excellent Set of Notes

With the basic questions in mind you have a head start on taking excellent notes. However, note taking is a skill and as is true with just about everything else in life, you only get better with practice. We provide a **Note-Taking Checklist** with specific actions you can take to get more of the content into your notes in a usable form. For many chapters, we provide additional note-taking tips that are specific to that particular chapter, e.g., suggestions about what to listen for or figures you need to review carefully before a particular lecture.

After Lecture

After each lecture we suggest you schedule two separate sessions: a brief note review session, and an **Intensive Study Session** (lasting at least an hour, maybe longer) during which you achieve mastery of that lecture's content.

Have You Checked Your Notes? Note taking is not confined just to lecture. Notes can almost always be improved upon after lecture. Good students know this and take the time after lecture to revise their notes. These are also the students who pull additional points on every midterm as their "paycheck." Specific suggestions in every chapter of the Study Guide will help you improve your lecture notes. Often these suggestions focus on adding visual material from the text or summarizing what was covered in an easy to remember form.

Intensive Study Session. The purpose of the intensive study session is to master the material you have just learned in lecture. Learning and cognitive psychology teaches us that passively reading new information is a very inefficient way to learn. You learn best if you spend your study time answering questions. With this method you read portions of the text as you need to find answers to questions. It's similar to the way I learn a new software package on my computer. I never read the software manual first. Instead I load a program and start using it. I go to the manual (or online help) when I need to find out if I can do something with the program, or to see how to do it. Applying this to the textbook, you will learn geology most efficiently by working backward from questions. To get you started, we suggest questions you can use this way. Often we suggest other guiding questions from *Understanding Earth* or the Web site. Simply stated: Do answer questions, don't "just read." You will learn and remember more science much faster.

Exam Prep

Doing well on college exams is mostly about organization. Effective study requires a systematic, orderly review. Most of your review time (about 70 percent) should be spent answering review questions. For each chapter of the text the guide includes a **Chapter Summary, Practice Exercises,** and **Review Questions.** Practice exercises, Web activities, and a set of multiple-choice questions are provided for every study session in every chapter. Using these is the fastest way to learn and the best way to remember the information you have learned. Study tips for preparing for midterm and final exams, helpful hints that will improve your test-taking skills on multiple-choice tests, and how to use your personal learning style to your advantage during exams may also be found in this section.

Brief Preview of the Study Guide for *Understanding Earth*   3

How to Study Geology

DURING LECTURE

Take an excellent set of notes.
- Note-Taking Checklist

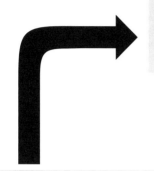

BEFORE LECTURE

Prepare for lecture: Arrive with an overview in mind
- Chapter preview
- Vital information from other chapters

RIGHT AFTER LECTURE
- Review notes.
- Fill in what you missed.
- Add visual material.
- Use the "Have You Checked Your Notes?" section of this guide.

AFTER LECTURE

Intensive study session
Master the key concepts.
- Web site activities and tools
- Practice Exercises and Review Questions

EXAM PREP

Begin review one week before the exam.
- Chapter summaries
- Practice Exercises and Review Questions
- Tips for Preparing for Geology Exams
- Eight-Day Study Plan in Appendix A

CHAPTER 2

Meet the Authors of *Understanding Earth*: How to Use Your Geology Textbook

As you begin your study of geology first take a minute or two to get acquainted with your text. First of all, why do you think it was written? Here is what your text authors have to say about why they wrote this book.

Geology fascinates and excites us. We wrote Understanding Earth *to help you discover for yourselves how interesting geology is in its own right and how important an understanding of geology has become for making decisions of public policy. What can we do to protect people and property from natural disasters, such as volcanoes, earthquakes, and landslides? How can we use the resources of Earth—coal and oil, minerals, water, and air—in ways that minimize damage to the environment? In the end, understanding Earth helps us understand how to preserve life on Earth.*

They add that people tend to enjoy what they do well. The authors designed their text to help you do well in your geology course. Many aids to learning are built into the text. Let's spend a minute or two talking about these aids and how you can use them.

Clues and Tools: "What's Important?"

One of the toughest things about taking an introductory course is that you typically have very little knowledge of the subject matter. Not only do you have a lot to learn, but you have a lot to learn about how to learn it. Where should you focus your attention? What skills and concepts should receive the bulk of your study energy? Aids to help you see what material is important are built into every page of *Understanding Earth*. However, you have to know where to look for these aids and how to use them. Here is a short list of learning aids in your new text and a few preliminary thoughts about how to maximize their usefulness.

✓ **Chapter Outline:** The authors begin each chapter with an outline of what will be covered. To use this tool you need to look actively for clues. Look at the outline for Chapter 1. Each item in the outline is a clue to what will be covered in the chapter. Try turning each item in the outline into a question. **Example:** Ask yourself "What is the scientific method?" as you read that section. Some items in the outline may surprise you. Pay particular attention to material in the outline you find surprising or puzzling in any way. Let's say you are surprised that a geology text has a section titled "The Origin of Our System of Planets." You can use your surprise or puzzlement to motivate yourself when you read that section. Frame your surprise into a question such as "Why in the world would a geology text called *Understanding **Earth*** be talking about other planets?" Read with that question in mind.

✓ **Chapter Summary:** At the end of each chapter a brief summary emphasizes the most important ideas of the chapter. This is a very useful resource when you are studying for an exam.

Time Saver Tip

Before reading a chapter of the text, read the Chapter Summary first, referring as you do so to the Chapter Outline. This will provide an overview that will greatly accelerate your reading and understanding of the material. **Hint:** Also use the **Chapter Preview** questions in this Study Guide. These questions are specially devised to help you focus on the most important material.

✓ **Key Terms** within the text are printed in bold. You can further speed up your reading time by being vigilant for key terms in bold. When you see a **boldfaced term** you know it is the most important concept in that paragraph. Focus on understanding that concept.

✓ **Photographs:** Geology is a very visual science. Pictures in the text are essential. Use the pictures as a "virtual field trip" to help you learn what particular rocks and formations look like. Pay particular attention to the photographs that are paired with schematic figures or diagrams as these can be a very useful visualization tool. Refer to these as you read.

Study Tip

Having a tough time getting yourself started with a study session? Use the art to motivate you. Start a chapter study session by scanning the images and captions to find material that interests you. Begin your reading with that material. This approach works like starting a fire; you generate a small spark of interest, and then fan it by bringing in new material.

✓ **Figures (Flowcharts and Tables):** Figures are even more important than photographs. Flowcharts such as the rock cycle in Figure 4.9 present the key concepts of each chapter. Pay careful attention to the arrows in flow models and ask yourself what drives the process. If you are a visual learner, you may even want to study the figures and sketches before you start to read. Then read text on an as-needed basis to clarify the figure.

✓ **Figure Stories:** In geology the story of a particular feature or area is often instructive. In each chapter of *Understanding Earth,* you will find a "figure story," which is an illustrated vignette that explains an important geologic process or principle. Figure stories often contain some data, affording you the opportunity to see how geologists interpreted that data and the kind of thinking they did to reach a conclusion.

When reading figure stories ask yourself how the story illustrates and develops what was said in the text. Try to understand the story, then test yourself by summarizing (out loud if you are an auditory learner) what you learned. Kinesthetic learners (see page 10) may find the figure stories particularly helpful because each provides a real-life example, similar to a lab or field experience. Visual learners will appreciate how the carefully chosen illustrations clarify the text.

✓ **Color** is used in the figures to provide you with clues about how each process works and what is involved. Look at Figure 1.11 in Chapter 1. What color is used to describe processes that utilize heat? What color depicts water? Cooling? Such cues can support learning at a subliminal level of awareness, particularly for visual learners, so be sure to pay close attention to the colors in the figures!

✓ **Clues to rock texture** are also cleverly built into the text. Look at Figure 4.4, which depicts igneous, sedimentary, and metamorphic rocks. The igneous rock is shown in shades of black with lots of dots and specks of varying size. This makes sense given that igneous rock is classified on the basis of the size of its crystals. Why are sediments and sedimentary rock depicted in earth tones with horizontal markings? Because sediments are formed

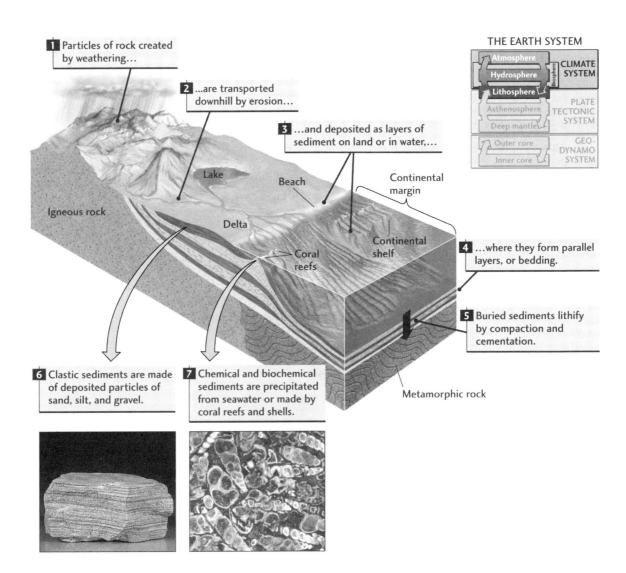

of earthy materials like sand and mud, and they are originally laid down in horizontal sheets and layers. Why is metamorphic rock depicted with a series of distorted and folded bars? Because metamorphic rock is created when other kinds of rock are subjected to high pressures and temperatures that may distort the rock into folds and its crystals into wavy, foliated bands. Thus, the figure speaks to us about what the rock looks like as well as the processes that produce it. Pay close attention to these visual learning clues.

Questions and Exercises to Help You Learn

✓ **Exercises** (end of chapter)
✓ **Thought Questions** (end of chapter)

Psychologists have firmly established the importance of actively learning by asking ourselves questions about the new information we are attempting to learn and then seeking the answers to these questions. When we formulate inquiries before we start reading new material, we tend to read in a more active manner because we are reading in search of specific answers. This is far more efficient than trying to read about the information you are attempting to *learn* in a passive manner (passive reading is OK for magazines, such as *Sports Illustrated,* but not for a college science textbook). Think of the end-of-chapter summaries and review questions as your own personal tutor. What does a good tutor do? She or he asks you questions to get you thinking about the material. Like a good tutor, the questions at the end of each chapter clue you in to what is important and encourage you to really master the material.

You can use both the **Exercises** and **Thought Questions** at the end of each chapter of *Understanding Earth* in three different ways. First, use them to *preview* the chapter. Skim quickly to get an overview of what you will be expected to know after you have read the chapter. Second, seek out answers to the questions given at the end of the chapter. Organize your study time around finding an answer to one or more exercises or questions. Then hunt for the answer in your text, using it as you would a reference book. Third, use the questions to review for exams. Quiz yourself by trying to answer the questions without using the text as your reference. Go back to your annotated text to check the accuracy of your answers. The best way to learn from a text is to use it as a reference to answer questions.

Text Marking Strategy for *Understanding Earth*: How to Mark and Annotate Your Text

You can make the 30 percent of the study time you spend reading the text even more efficient by skillful text marking. Marking your text as you read is a vital part of learning the material. This may be particularly important if you are a kinesthetic learner. The action of marking will aid your kinesthetic memory. If you are a visual learner, you will tend to remember marked sections during exams.

Good marking involves both thinking and planning. First, you think about what is important. Then plan for exam review, deciding which of the marking techniques in the table on page 8 will best help you when you return to the chapter during exam review. Only then should you mark. The authors have already marked **key terms** in **bold** letters. You will want to add underlines, numbers, and annotations in the margin. The table suggests how and when to use each kind of mark.

Marking the text will aid kinesthetic learning. *Good* text marking will aid visual memory—so use caution when marking up your text. *Mark carefully* and *selectively.* If you mess up the text with excessive messy markings, you may interfere with some of the good memory aids that are built into the careful design of the text. Remember: A well-marked text will make exam review far more efficient because the marks will focus your attention on the key material you need to find and review. If you are worried about the impact of marking on your textbook's resale value, be sure to ask your bookstore. Many bookstores have policies that encourage effective text marking.

Useful Text Markings for Geology	Marking Tips
<u>Underline important points</u>. Underline to highlight the <u>key ideas</u>. Pay attention to the <u>key terms</u> that are underlined for you in the text.	1. Read before you mark. To avoid overusing underlining set your pencil down on the table as you begin to read a paragraph. Don't let yourself pick up the pencil until you have read the entire paragraph. Stop and think. Then, underline what you consider to be the key point(s). 2. Be selective. Underline brief, but meaningful <u>phrases</u> or <u>key words</u> that will trigger your memory. Mark just enough so you can <u>review without rereading</u> the paragraph. 3. Use colorful markings if it helps you remember.
Annotate the Margin Write a brief summary statement of key geology processes in your own words in the text margin. Use these annotations to facilitate exam review.	1. Use your own words. Putting ideas into your own words is a powerful learning strategy. 2. Be neat. This takes time. However, it will pay off later when you review, because your annotation will be legible and easily read. 3. Organize your annotations into categories. Grouping ideas into categories or bulleted lists makes them easier to remember. 4. Use annotations during exam review. Avoid rereading the text word for word. This will save you a lot of time.
Useful marks to use in the text margin:	Circled numbers in the margin indicate sequences, such as the processes of the rock cycle. **Question Mark ?** Put a question mark in the margin to remind yourself to ask your instructor about a point you do not understand. **Asterisks ***** Use asterisks to mark ideas of special importance. Use asterisks sparingly. Save them for the two or three most vital ideas in the entire chapter. **TQ (test question)** Use TQ to note material in the chapter that you know will be covered on the exam. This will remind you to pay particular attention to these items when you review.

TRY THIS NOW

Turn to the chapter you are currently studying. Experiment with one or more of the following.

1. Try underlining. Choose a section you consider to be important, key material. Read the passage several times until you are sure you understand it. Then mark the passage following the directions in the table.

2. Annotate a geological process. Choose a section that contains a geological process that you think is important enough to be covered on the next exam. Annotate the process in the margin in a manner that will help you remember it. Use the directions in the table as a guideline.

3. Try some additional marking. Where would circled numbers be helpful in this chapter? Insert these numbers on the pages. What material do you need to discuss with your instructor? Mark with a question mark.

4. Figure out what is the most important idea in the entire chapter. Put three asterisks (***) and a TQ beside that section to ensure that you master it before the exam.

CHAPTER 3

How to Be Successful in Geology (and Just About Any Other Challenging Course)

Academic success is largely a matter of strategy. If you manage your time well and study strategically (rather than haphazardly), you will be successful. In this chapter, we discuss successful studying strategies for your geology course. The best way to use this chapter is to read each strategy and then try it out. We recommend that you read one section of this chapter each day in the beginning of a new semester. Try the strategy in that section before reading further. The success strategies that will be covered in this introduction are

- Learning style
- Chapter preview
- Note taking
- Note review
- Exam preparation

Many other learning strategies and hints about learning specific material are provided throughout this Study Guide. Look for them and try out as many of the strategies as possible. Learning to study in this way initially takes an investment of time, but you will find that this is one investment that pays off. Studying strategies make you more efficient so that you can learn more in less time. More important, you enjoy your learning more because you know exactly what you are doing!

Customize Each Strategy to Your Learning Style

Each of us learns differently. Successful people in any activity tend to be those who find a way to express their own unique, individual talents in that activity. Most of us worry too much about competing. We would do better to seek a way of using our best skills and talents. It is our individuality that makes us stand out from the pack and become a leader. It is our individuality that leads us to be successful academically. We recommend that you spend some time thinking about your individual style, your strengths as a learner, and how you prefer to learn. Here's a simple beginning. Which mode of learning do you *prefer to use* and *use best*?

- ☐ **Visual Learning** Visual learners learn by seeing. They often have a good memory for pictures and even words of text.
- ☐ **Auditory Learning** Auditory learners learn by listening. They are good listeners and remember best by just listening.
- ☐ **Kinesthetic Learning** Kinesthetic learners learn by moving. They learn and remember best when they get to practice an activity.

TRY THIS NOW

Put a 1 in the box next to your strongest and preferred mode of learning, a 2 for you next strongest mode, and a 3 for the mode you think you use least well and least prefer to use. **Hint:** Usually, the mode we prefer is the one we use best. If you are not sure which mode you prefer, try taking the Learning Style Inventory on the University of Arizona Learning Center Web site: www.ulc.arizona.edu.

Throughout this Study Guide we will make suggestions about how particular strategies *may* work best for those who prefer one of the aforementioned learning styles.

Make Geology Lecture a High-Priority Activity!

This is step one in any strategic approach to course success. Your geology course centers on lecture. Geology is a very content-intensive subject. There's so much to learn that instructors have to make some difficult decisions about what to cover. Sometimes faculty in a geology department try to address this problem by mapping out required content, perhaps even creating a core syllabus for instructors to follow. Even then instructors can and do manage to create very different courses. In our conversations with other instructors we have been quite impressed with the degree to which teaching approaches differ. Two geology instructors may use the same text and perhaps even the same basic lecture outline yet still teach two very different courses *with very different midterm exam questions*. Therefore, attending lecture is fundamental to success. Most students who fail geology do so because they fail to attend lecture. Just attending will put you far ahead of those who fail.

However, just attending is not enough to ensure that you get an A in geology. Research shows that few of us have the attention span to listen actively and take good notes for an entire hour. The average note taker gets most of the main points during the first 10 minutes of the lecture. Then attention wanders, confusion sets in, and less and less of what is said gets into the notes. Indeed, most students get less than 15 percent of the points that are covered in the final 10 minutes of the lecture. This is doubly unfortunate because, inevitably, the lecture gets to the heart of the topic and the material you will certainly be tested on toward the end of the lecture.

What you actually do in lecture is fundamental to success in geology. You need to be a strategist. You need to target your approach to your geology course on getting the most possible out of lecture. That is our best advice, and we have adhered to it rigorously in constructing this guide. The entire Study Guide is organized around lecture. Every chapter begins with what you do *before lecture* for that particular chapter topic. Next comes a section with tips on what to do *during lecture,* then a checklist to help you improve your notes *after lecture,* and finally several exercises and review questions to help you with *exam prep* before your midterm.

In this chapter we present general strategies you can use every day of your course to ensure success. Here again the chapter is organized around lecture.

Before Lecture

Strategy: Preview the Chapter Before Going to the Lecture

To preview a chapter use the **Chapter Preview** questions. These questions are provided at the beginning of each chapter in the Study Guide. Read the question; then skim the chapter to find the relevant material. Feel free to annotate the brief answers provided for each question. The idea is to go to lecture with these questions and a general idea of the answers in mind.

Why Preview? Introductory courses can be difficult. There are lots of new terms, ideas, and skills during a first-semester geology course. Which ideas are "important"? How do you focus your effort in this new and seemingly strange terrain?

> *Picture a house under construction. Imagine further that the contractor was very careless and neglected to construct whole sections of the frame. Finishing those areas lacking a frame would be impossible. You can't tack siding to thin air! This metaphor describes your geology lecture. You are the contractor. You need to arrive at class with an overview of the lecture already in mind. An overview means you have already identified what geological processes will be explained and what key questions the lecture will answer. Lacking a frame of key questions is like building a house with no supporting structure. You will have nothing on which to hang the lecturer's main points (information). Without the main points the details are meaningless. As the lecture progresses you are likely to feel increasingly confused and bored. By the time the lecturer gets to the most important material you may be completely lost.*

Where do you get the overview? The answer is surprisingly simple. You just need to spend a few minutes before lecture previewing the chapter that will be covered. Previewing is the method by which you generate and master a framework for listening. Here's how to do it.

Step 1. Read the preview questions. You will find **Chapter Preview** questions at the beginning of each chapter in your Study Guide.

Step 2. Skim the text chapter for answers. Quickly find and read only the material in *Understanding Earth* you need to understand and answer the question. Don't bog down on details. Don't try for complete understanding. Don't read the whole chapter. Your goal is to gain just a general understanding of each question. Do use the **brief answers** provided in the Study Guide to cue your reading.

Step 3. Memorize the questions and "brief answers." You may find you need to add a few notes to our brief answers in the Study Guide. Annotate our answer just enough to ensure that the answer makes sense to you. Annotate in a way that will ensure that you memorize the questions.

Before you go to lecture always be sure to spend some time previewing. You will find that as little as 10 to 15 minutes of time spent previewing can make a big difference in how much you understand of the lecture. With the key points already in mind, you can focus in lecture on understanding the details. This, in turn, will help ensure that you get an excellent set of notes.

TRY THIS NOW

Move immediately to the chapter that will be covered in your next lecture. Preview that chapter. Then return to read the next strategy.

During Lecture
Strategy: Note-Taking Checklist

Your basic goal during lecture is to get an excellent set of notes. Those notes should contain in-depth answers to the **Chapter Preview** questions. To avoid getting lost in details, keep the big picture in mind. Often this is best accomplished by having a copy of the **Chapter Preview** questions in front of you. Better yet, develop preliminary answers to the questions and commit them to memory before the lecture.

Another way to keep the big picture in mind is to bring a copy of the key figure or flowchart to lecture to refer to. In many chapters, we suggest which figure you should have handy.

During some classes your lecturer may show rock formations and pass around specimen rocks. You will get more out of these if you sit close to the front row where you can see the sample rocks as the lecturer discusses them. Remember to focus carefully on clues provided for recognizing a particular sample in the field. Focus in particular on the texture of samples and learn to recognize differences. For example, you will learn the difference between a fine-textured volcanic rock and a coarse-textured plutonic rock.

As you listen to lecture identify questions you need to ask to understand the material. Try to formulate at least one good question you can ask during every lecture.

Of course, the most important task during lecture is taking a good set of notes. Note taking is not easy. It is a skill that improves with practice. Here are a few tips that will help ensure that you take an excellent set of notes.

Note-Taking Checklist

- ☐ Organize your notes in a three-ring binder so that you can easily reorganize them.

- ☐ Save space in your notes to add important visual material (flowcharts, simple sketches, comparison charts) after class. To make this easy, employ a **double-column or double-page note-taking format.** Take notes on the right-hand page or column. Save the left-facing page as your "sketch page."

- ☐ Sit near the front of the room. This has nothing to do with being a teacher's pet—this is simply for hearing and seeing better.

- ☐ Date each day's notes for locating material later.
- ☐ Take notes in a format that makes the main topics and concepts easy to identify. Some students accomplish this by taking notes in outline format. But many other approaches are possible. Visual learners may find it helpful to highlight main points (after class) in color. Another good approach would be to use the questions we provide under Before Lecture: Chapter Preview as headers. These can be added during class or during your after-class review session.
- ☐ Keep the preview questions in front of you during lecture. Be sure to leave class with good answers to each of the three preview questions.
- ☐ Indicate areas where you need to do follow-up work with your text, instructor, or tutor by placing a question mark in the margin next to that material.
- ☐ Indicate possible test questions by putting a TQ (test question) in the left margin where you can easily see it.
- ☐ Write assignments in the left column where you can easily find them. After class enter the due date in your personal planner or calendar.

TRY THIS NOW

Take the note-taking checklist with you to your next lecture. Try to follow all the suggestions. After lecture, review your notes and check off each point that you actually followed. Then return to read the next strategy.

Hint: Why not photocopy the checklist and use it for every lecture until all the activities have become habits? Used this way the checklist becomes a visual record of your progress as a skillful note taker.

After Lecture

Review Your Notes Immediately

Good note taking continues after the lecture is over. Right after lecture, while the material is fresh in your mind, review your notes. Review to be sure you understand the key points and wrote them down in a form that will be easy to review later.

Don't postpone this activity. The best time is right after lecture before you go to your next class or activity. Learning experts tell us that most of us forget 80 percent of what we hear in a lecture by the following day. But if you review right after the lecture, there will have been no interruptions. Much of what was said will still be in your short-term memory. If you missed something, you can probably remember it and add it to your notes. The basic idea of reviewing your notes is to fill in what you missed and add helpful visual material from the text. Use the following checklist as a guide.

Check Your Notes: Have You . . .

- ☐ written legible notes? (Rewrite your notes if necessary.)
- ☐ identified the important points clearly? (You should have headers in your notes for each of the questions in the **Chapter Preview**.)
- ☐ filled in the holes (missing material) while the lecture is still fresh in your short-term memory?
- ☐ marked areas where you don't remember what was said for a follow-up session with your instructor, tutor, or study partner?
- ☐ indicated possible test questions (TQ) in the margin?
- ☐ added visual material?
- ☐ reworked notes into a form that is efficient for your learning style?
- ☐ created a brief "big picture" overview of this lecture (using a sketch or written outline)?

TRY THIS NOW

After your next lecture review your notes and improve them using the steps in the checklist as a guide. Check off each step as you complete it. Then return to read the next strategy. **Hint:** Again, consider photocopying the checklist so that you can use it repeatedly.

Intensive Study Session

Ask yourself questions as you study; then answer the questions.

You should schedule at least one hour after each lecture for intensive study. This can occur anytime before the next lecture. (Short-term memory is no longer a problem because you have completed a note review and have a good set of notes.)

Why do you need an intensive study session? Think about the house/construction example that we mentioned earlier. Just as you need a frame before you can add the siding, so you construct a frame of questions before each lecture. During lecture, you add the siding by attaching details and ideas to the questions. After lecture, master the ideas during an intensive study session to construct the first story. This geology course is a skyscraper with 23 floors (one for each chapter). Each chapter supports those above it. If you don't completely master this chapter, the next one will be more difficult.

Mastery is not gained by just reading the text. Mastery occurs as the result of asking yourself questions (and answering them). To help you, we provide **Practice Exercises and Review Questions** for every chapter of the text. This interactive learning material is specifically designed to help you master the key concepts of each chapter. The greater the number of these exercises you can work into your intensive study sessions early in the course, the easier subsequent chapters will be. Plan to spend the majority of your study time (70 percent) working on these exercises.

Use your text as a reference. Read it as needed in order to answer questions and master material. When you read, read efficiently. Read with purpose and read to find answers to questions. **Hint:** To use your text effectively you must know it like a book. Carefully read the **Meet Your Authors** in this guide.

TRY THIS NOW

Each chapter of this Study Guide contains a section titled **Intensive Study Session.** Turn now to the chapter you are currently working on and try out the **Intensive Study Session** material for that chapter. Do enough to get a feeling for how the **Intensive Study Session** works. Then return and read the following section.

Exam Prep

Materials in this section are most useful during your preparation for midterm and final examinations. For optimal performance midterm preparation should begin about eight days prior to the exam (see the **Eight-Day Study Plan**). The basic idea is a systematic review of material divided into short study sessions.

Tips for Preparing for Geology Exams

✓ Use the clues your instructor has provided in lecture about what is important. Even when a department agrees on a common core of material (a very rare occurrence), each instructor carves out a course that is unique, has a particular character or flavor, and has distinct areas of emphasis. Your instructor is the ultimate guide on what is important.

✓ Be sure you know the format of the exam. Multiple choice? True-false? Essay? Thought problems?

✓ Review your notes for material marked TQ (test question) in the textbook.

✓ Ask your instructor if exams are available from the previous semester. Review these to check the format of questions, what areas of content are stressed, and what types of problem solving are included. Don't make the mistake of assuming that the same questions will be asked this semester.

✓ Be sure to attend review sessions if they are offered.

✓ If your class has tutors, preceptors, supplemental instruction leaders, or other peer helpers who have taken the course, ask for their suggestions about preparing for the exam.

✓ Once you are clear about the nature of the exam begin your review. Conduct your review in an orderly, systematic manner that ensures a focused review of all the important material. The **Eight-Day Study Plan** (refer to **Appendix A**) is a good model for an orderly review.

Test Taking

In every college exam there are a number of students who know the material yet fail the exam because they get anxious, panic, and freeze up. I have found over the years that the best way to avoid or overcome test anxiety is by coming to the exam well prepared and confident and by working strategically on the test. Exam prep will not be a problem if you use the materials provided for this specific purpose within this Study Guide. In this section, we will suggest strategies to try out during your exam. You may want to return to this section a day or so prior to your exam.

Test-Taking Tips

Test Taking and Learning Style

Visual Learners
- Use written directions.
- When you get stuck on an item, close your eyes and picture flowcharts, pictures, field experiences, or text.

Auditory Learners
- Pay attention to verbal directions.
- Repeat written directions quietly to yourself (moving your lips should be enough).
- If you get stuck, remember your lecturer's voice covering this section.

Kinesthetic Learners
There are a variety of things kinesthetic learners find helpful when they get stuck on a test item. Try some of these.
- If you get stuck move in your chair or tap your foot to trigger your memory.
- Feel yourself doing a lab procedure.
- Sketch a flowchart to unlock the memory of a process.

Many exams are in multiple-choice format. Here are some tips to maximize your performance on multiple-choice questions.

Test-Taking Tips

Top Ten Tips for Taking a Multiple-Choice Exam

10. Answer the questions you know first. Mark items where you get stuck. Come back to harder questions later. Often you will find the answer you are looking for embedded in another much easier question.

9. Try to answer the item without looking at the options. Then check to see if your answer matches any of the options.

8. Eliminate the distracters. Treat each alternative as a true-false item. If "false," eliminate it.

7. Use common sense. Reasoning is more reliable than memory.

6. Underline key words in the stem. This is good to try when you are stuck. It may help you focus on what question is really being asked.

5. Look for similar alternatives. It is likely that one of them is correct.

4. Answer all questions. Unless points are being subtracted for wrong answers (rare), it pays to guess when you're not sure. Research indicates that items with the most words in the middle of the list are often the correct items. Be cautious, as your professor may have read the research too!

3. Do not change answers. Particularly when you are guessing, your first guess is often correct. Change answers only when you have a clear reason for doing so.

2. If the first item is correct, check the last. If it says "all (or none) of the above," you obviously need to read the other alternatives carefully. Missing an "all of the above" item is one of the most common errors on a multiple-choice exam. It is easy to read carelessly when you are anxious.

1. Read the directions before you begin!

Final Exam Week

Each semester in one short week you take an exam in each and every course. Most of the exams are comprehensive finals that cover the entire semester. Dealing with finals week successfully is a major challenge. Here are some tips that will ensure that you do your best work during finals week.

Tips for Surviving Finals Week

Be organized and systematic. Use the **Final Exam Prep Worksheet (Appendix B)** to help you get organized for finals. Use the **Eight-Day Study Plan (Appendix A)** for every course where the final exam will be an important factor in determining your grade.

Stick to priorities. Say no to distractions.

Build in moments of relaxation. Take regular short breaks, exercise, and get enough sleep.

Be confident. By now you have built up a good set of study habits. You are a competent learner.

Organizing for final exams can be quite a challenge! As with any big project, devoting some organizing time up front will pay big dividends. Modify the suggestions in the **Final Exam Prep Worksheet** to fit your personal situation and needs.

We wish you success!

CHAPTER 1

Building a Planet

Civilization exists by geological consent. . . .
subject to change without notice.

—WILL DURANT

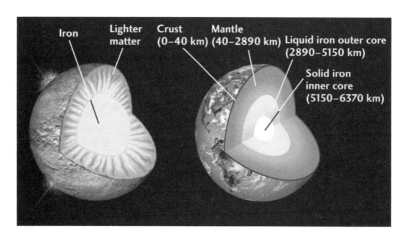

Figure 1.6. The differentiation of early Earth resulted in a zoned planet with a dense iron core, a crust of light rock, and a residual mantle between them.

Before Lecture

Chapter Preview

- **How do geologists study Earth?**
  Brief answer: Using the scientific method. Figure 1.1 is an excellent preview to the process.

- **How did our solar system form?**
  Brief answer: It accreted from gas and dust about 4.5 billion years ago. Figure 1.3 shows this process beautifully. Look it over before lecture.
- **How did the Earth form and change over time?**
  Brief answer: The Earth's core, mantle, crust, oceans, and atmosphere evolved as the interior of the planet heated up, melted, and differentiated. Study Figure 1.6.
- **What are the major components of the Earth system?**
  Brief answer: Table 1.2 and Figure 1.10 illustrate how Earth is a system of interacting components, including the atmosphere, hydrosphere, biosphere, and interior.
- **What were some key events in the history of the Earth?**
  Brief answer: Figure 1.12 presents the highlights. Focus on the connections between geologic and life history.

During Lecture

This will be your very first lecture for this course. Get off to a good start. Arrive 10 minutes early. Find the best seat in the house—close to the front of the room where you can hear the lecture and see the slides. This section is referred to by some as the "A section." Quite apt, there really is a correlation between where you sit in a lecture hall and the grade you are likely to receive. Test this out. Try sitting in different locations during the first week. Notice where it is easiest to concentrate and where it is not. In many lecture halls the back row is the worst place to be. It is the location chosen by students who arrive late and leave early. It may even be noisy. Always feel free to change locations if your view is impeded by projection equipment or whispered conversations of other students. Talk to your instructor if necessary.

OK, you've got a good seat. What can you do while you wait for the lecture to begin?

- **First, motivate yourself to want to listen to this lecture.** Open your text to Chapter 1 and thumb through the chapter. Take a good look at the photos and figures. Look for material that interests you. Chapter 1 is loaded with interesting visual material so this should be easy to do. Ask yourself what you would like to know about this chapter. What would you ask your teacher if this were a one-on-one tutorial? Finally, try to think of some experience you have had that relates to this chapter. Maybe you visited a planetarium, read a catchy version of the formation of the universe, such as *Cosmic Comics* by Italio Calvino, or saw a volcano on your last vacation. Actively look for experiences to hook your interest and personally connect with course information. Notice how the more you look at the chapter, the more you let yourself think about the pictures and the more your interest builds. Five minutes should be enough to get your "motivational engine" tuned up and humming.
- **Second, prepare your mind for learning.** A master football player warms up by stretching, running, and passing the football. A master learner warms up by focusing his or her attention on what will be covered during lecture. Spend a minute or two looking over the **Chapter Preview** questions. Try to anticipate how these questions will be answered in lecture. If you have time, read the **Chapter Summary** for Chapter 1. When you preview, the goal is not to learn the material but merely to formulate questions that you expect will be answered during the lecture. With questions in mind you are ready to take notes.

LearningTip

Do a "learning warm-up" before every lecture. Arrive 10 minutes early. "Psych" up. Prepare your mind for learning. Anticipate questions that will be answered in lecture.

The man who is afraid of asking is ashamed of learning.
—DANISH PROVERB

After Lecture
Review Notes
The perfect time to review your notes is right after lecture, while the material is still fresh in your mind. Review to be sure you got all the key points and wrote them down in a form that will be readable later. As you review you can also polish your notes by adding useful visual material and a summary.

Check your notes: Have you...

☐ added a simple sketch or two to clarify the key points? Hint: Try sketching your personal version of how convection drives plate tectonics based on Figure 1.11. Make it simple, something you will easily picture and remember.

☐ written a brief (one-paragraph) summary of the most important concept you learned from this lecture? Feel free to draw on your notes and figures in your text as needed. Reviewing preview questions may help.

You will spend some of your study time reading *Understanding Earth,* particularly sections of the text that are emphasized in lecture. The text is loaded with tools and clues to help you learn. You will find our suggestions about how to take advantage of these learning aids in Meet Your Authors (Part I, Chapter 2) of this Study Guide.

Web Site Study Resources
http://www.whfreeman.com/understandingearth

Check out the web site to get an idea of available study aids. You will find **Concept Self-Checker, Web Review Questions, Graded Quizzing, Online Review Exercises,** and **Flashcards** (to help you learn new terms). The *Online Review Exercise: Identify the Plate Boundaries* is a good review of plate tectonics, to be used for Chapters 1 and 2.

Exam Prep
Materials in this section are most useful during your preparation for midterm and final examinations.

The following **Chapter Summary**, and **Practice Exercises and Review Questions** should simplify your chapter review. Read the **Chapter Summary** to begin your session. It provides a helpful overview that should refresh your memory.

Next, work on the **Practice Exercises and Review Questions.** In order to determine how you stand on mastery of this chapter, complete the exercises and questions just as you would a midterm. After you answer the questions score them. Finally, review any question that you missed. Identify and correct the misconception(s) that resulted in your answering the question incorrectly.

Chapter Summary
How do geologists study Earth?

- Field and lab observations, experiments, and the human creative process help geoscientists to formulate hypotheses (models) for how the Earth works and its history. A hypothesis is a tentative explanation which can help focus attention on plausible features and relationships of a working model. If a hypothesis is eventually confirmed by a large body of data, it may be elevated to a theory. Theories are abandoned when subsequent investigations show them to be false. Confidence grows in those theories that withstand repeated tests and are able to predict the results of new experiments.

How did our solar system form?

- Our solar system probably formed when a cloud of interstellar gas and dust condensed about 4.5 billion years ago. The planets vary in chemical composition in accordance with their distance from the Sun and with their size.

How did the Earth form and change over time?

- Earth probably grew by accretion of colliding chunks of matter. Very early after the Earth formed, it is thought that our Moon formed from material ejected from the Earth by the impact of a giant meteorite.
- Heat generated from the Moon-forming impact and the decay of radioactive elements probably caused much of the Earth to melt. Melting allowed iron and other dense matter to sink towards the Earth's center and form the core. Lower density (lighter) matter floated upward to form the mantle and crust. Release of trapped gases (mostly water) from within the Earth gave rise to the oceans and an early atmosphere. In this way, the Earth was transformed into a differentiated planet with chemically distinct zones: an iron core; a mantle of mostly magnesium, iron, silicon, and oxygen; and a crust rich in oxygen, silicon, aluminum, calcium, potassium, sodium, and radioactive elements.

What are the major components of the Earth system?

- As the Earth cooled, an outer relatively rigid shell called the lithosphere formed on top of a hotter and softer asthenosphere. Volatiles trapped within the mantle escaped through volcanoes to form Earth's atmosphere and oceans. Life evolved and developed the capability of extracting carbon dioxide out of surface environments and releasing oxygen gas as a by-product of photosynthesis.
  The interaction of major components of Earth's system continues to this day.

What were some key events in the history of the Earth?

- Earth was formed 4.5 billion years ago.
- Earliest life forms are found in rocks about 3.5 million years old.
- By 2.5 billion years ago, photosynthesis by early plant life produced increasing levels of oxygen in Earth's atmosphere.
- Only one-half billion years ago animals appeared, diversifying rapidly in an explosion of evolution.
- Life history is marked by periodic mass extinctions. About 65 million years dinosaurs were killed off in an extinction event caused by a large bolide impact.
- Our species arrived 120,000 years ago—a tiny fraction of the 4.5-billion-year history of the Earth.

Practice Exercises

Answers and explanations are provided at the end of the Study Guide.

Exercise 1: The Evolving Early Earth

Fill in the blanks in the flowchart below which characterizes a popular hypothesis for the early history of the Earth after the Sun and planets had formed (refer to adjacent flowchart).

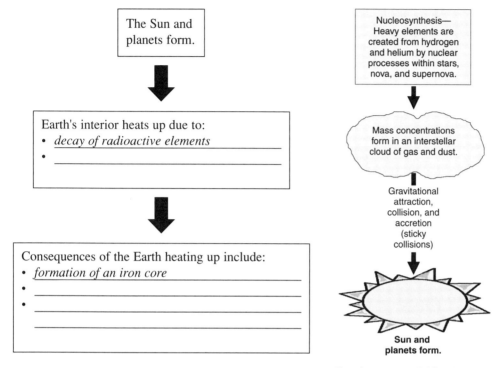

Flowchart—a model for the formation of the solar system.

Review Questions

Answers and explanations are provided at the end of the Study Guide.

1. Of the following statements regarding the scientific method, which one is true?
   A. A hypothesis that has withstood many scientific tests is called a theory.
   B. The outcome of scientific experiments cannot be predicted by a hypothesis.
   C. For a hypothesis to be accepted, it must be agreed upon by more than one scientist.
   D. After a theory is proven to be true, it may not be discarded.

   Hint: Figure 1.1 is an excellent overview of the scientific method.

2. During the formation of our solar system, what was the process that caused dust and condensing material to accrete into planetesimals?
   A. Nuclear fusion.
   B. Rapid spin of the proto-Sun.
   C. Heating of gases.
   D. Gravitational attraction and material collisions.

3. A major source of internal heat in the Earth today is
   A. ocean tides.
   B. radioactivity.
   C. solar energy.
   D. volcanoes.

Test-Taking Tip

When taking a test be alert to items that give away the answers to other questions.
Example: Question 15 may provide a hint for answering Question 4.

4. Which of the following processes is thought responsible for the formation of the present Earth atmosphere?
   A. Chemical breakdown of minerals.
   B. Capture of oxygen and nitrogen from space.
   C. Both degassing of the Earth's interior and oxygen liberated by photosynthesis.
   D. Degassing of the Earth's interior.

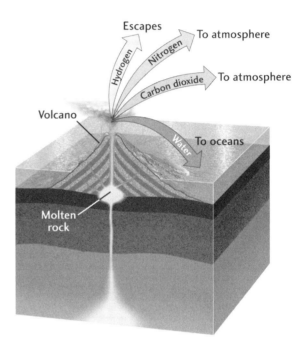

Figure 1.8. Early volcanic activity contributed enormous amounts of water vapor, carbon dioxide, and other gases to the atmosphere and oceans and solid materials to the continents. Photosynthesis by microorganisms removed carbon dioxide and added oxygen to the primitive atmosphere. Hydrogen, because it is light, escaped into space.

5. How old is the Earth estimated to be based on radiometric dates of lunar rock samples and meteorites?
   A. 45 million years old.
   B. 4.5 billion years old.
   C. 10 billion years old.
   D. 100 billion years old.

6. Geochemists have determined that differences in the distribution of chemical elements in the present Earth's crust, mantle, and core—in contrast to the distribution of chemical elements in the initial solar nebula—may be due in part to
   A. inhomogeneities within the initial interstellar dust cloud.
   B. early melting and differentiation of materials of varying density.
   C. cosmic ray bombardment over 4.5 billion years.
   D. nuclear synthesis within our Sun.

7. Why do geoscientists think tectonic plates move across the Earth's surface?
   A. Centrifugal force of Earth's rotation spins plates across the Earth's surface.
   B. Volcanic eruptions on the seafloor push tectonic plates apart.
   C. Tidal forces drive plate motion.
   D. Movement of the plates is the surface manifestation of convection in the mantle.

8. Within our solar system one big difference between the inner and outer planets besides size and position relative to the Sun is their
   A. density.
   B. color.
   C. rate of formation.
   D. time of formation.

9. Earth's Moon is not thought to have formed by
   A. capture of a large planetary object traveling past the Earth.
   B. accretion similar to the planets.
   C. material ejected from Earth by volcanic eruptions.
   D. the impact of a Mars-sized object on Earth very early in Earth's history.

10. The nebular origin of our solar system is characterized as
    A. a proven fact.
    B. a theory.
    C. a hypothesis.
    D. pure guesswork.

11. What valuable purpose does a scientific hypothesis serve?
    A. It reveals the absolute truth.
    B. It provides a basis for testable predictions.
    C. It can never be changed.
    D. It is a complete representation of the real world.

12. Bombardment from space may be disastrous for life but is also an essential process in the history of a planet. Why?
    A. Bombardment is how a planet grows and residual heat from impacts may help to create a dynamic planet.
    B. The oceans and atmosphere formed on Earth as a result of bombardment.
    C. Impacts drive plate tectonics.
    D. Impacts keep planets from getting too big.

13. The impacting object that caused the extinction of the dinosaurs 65 million years ago is estimated to have had a radius of about
    A. 100 meters.
    B. 1 kilometer.
    C. 10 kilometers.
    D. 100 kilometers.

14. Convection transfers heat (see Figure 1.11) by the physical circulation of hot and cold matter. How does it work?
    A. Heated matter rises under the force of buoyancy because it is less dense.
    B. Hot matter within the mantle sinks because it is denser.
    C. Cold matter rises under the force of buoyancy because it is less dense.
    D. Bombardment acts to stir up the mantle and drive convection.

15. Oxygen released into Earth's atmosphere by photosynthesis is vital for our existence not only because we need it to breathe but also because
    A. it is essentially for all life on Earth.
    B. it combines with free hydrogen gas in our atmosphere to produce water.
    C. life would not have evolved without oxygen.
    D. it forms an ozone layer in the upper atmosphere that protects us from UV radiation.

It is in the stars.
The stars above us, govern our condition.

—W. Shakespeare
King Lear, IV, iii

Figure 1.4. The solar system. The figure shows the relative sizes of the planets and the asteroid belt separating the inner and outer planets.

CHAPTER 2

Plate Tectonics: The Unifying Theory

From time to time in the history of science, a fundamental concept appears that unifies a field of study by pulling together diverse theories and explaining a large body of observations. Such a concept in physics is the theory of relativity; in chemistry, the nature of the chemical bond; in biology, DNA; in astronomy, the Big Bang; and in geology, plate tectonics.

—UNDERSTANDING EARTH, FRANK PRESS AND RAYMOND SIEVER

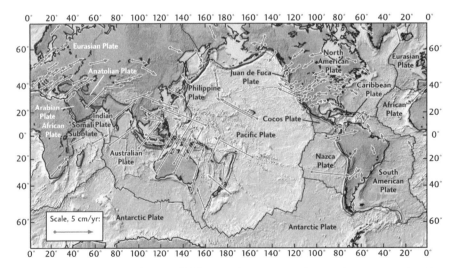

Figure 2.13. Global Positioning System (GPS) is used to measure plate motions at many locations on the Earth. The velocities shown here are determined for stations that continuously record GPS data.

Before Lecture

This is a particularly important chapter. Plate tectonics is the major concept that guides modern geology. Before you attend lecture be sure to spend some time previewing the chapter. Previewing will greatly increase your understanding of the lecture (see the Chapter Preview in Part I, Chapter 3: **How to Be Successful in Geology**). For an efficient preview use the following questions.

Chapter Preview

This chapter contains figures and photos that are extraordinarily helpful. The goal is to learn how it is that the continents and ocean plates can float about the Earth. Focus your Chapter 2 study time on examining these fascinating figures and photos of some of Earth's most amazing features.

- **What is the theory of plate tectonics?**
  Brief answer: The theory of plate tectonics describes the movement of lithospheric plates and the forces acting between them. It also explains the distribution of many large-scale features that result from movements at plate boundaries: mountain chains, earthquakes, volcanoes, topography of the seafloor, and distribution of rock assemblages and fossils. Refer to Figure 2.5 (Earth's plates today) on page 29.

- **What are some of the geologic characteristics of plate boundaries?**
  Brief answer: Volcanoes and earthquake activity are concentrated along plate boundaries. Where divergent plate boundaries are exposed on land, subsiding basins and volcanism are typical. Mountain chains form along convergent and transform plate boundaries. It is important to visualize each major type of plate boundary. Use the following figures in the text to do so: Figures 2.6, 2.7, and 2.8 for divergent plate boundaries; Figure 2.9 for convergent plate boundaries; and Figure 2.10 for a transform boundary that cuts continental crust. If you have actually seen or lived near the San Andreas, a volcano, or a rift valley, such as the Rio Grande in New Mexico, it would be more meaningful to start with that feature.

- **How can the age of the seafloor be determined?**
  Brief answer: A pivotal discovery in the history of plate tectonics was the determination of the age of the seafloor using magnetic anomalies. Figure Story 2.11 shows how this was accomplished.

- **How is the history of plate movement reconstructed?**
  Brief answer: Transform boundaries indicate the directions of relative plate movement and seafloor isochrons reveal the positions of divergent boundaries in earlier times. Refer to Figures 2.14 and 2.15.

- **What drives plate tectonics?**
  Brief answer: Earth's internal heat creates convective currents (flow of rock material from hotter to cooler areas) in the mantle. Convection, the force of gravity, and the existence of an asthenosphere are all important factors in any explanation of plate movement. Review Figure 1.11 to determine how convection works. Then, look at Figure 2.16 to see how convection coupled with the push and pull of gravity may drive plate movements.

Vital Information from Other Chapters

Review Figure 1.11 in Chapter 1 to remind yourself how convection drives plate tectonics.

Web Site and CD Preview

http://www.whfreeman.com/understandingearth
Identify the Earth's Plates and *Identify the Plate Boundaries* are **Online Review Exercises** worth completing before your first lecture on this topic.

During Lecture

One goal for lecture should be to leave class with a good set of answers to the preview questions.

- To avoid getting lost in details keep the "big picture" in mind. Chapter 2 tells the story of plate tectonics—Earth's moving plates, how the plates move, and the geological features associated with converging and diverging plate boundaries. Plate tectonics underlies and explains much about modern geology. In that sense, this chapter provides a preview of your entire geology course.
- Focus on understanding Figures 2.6 and 2.9. It will be helpful to have these figures handy during lecture. Annotate the text figures with comments made by your instructor.
- Want ideas on taking a good set of notes? If you haven't already done so, read our discussion of note taking in **How to Be Successful in Geology** (Chapter 3 in Part I of this Study Guide). **Hint:** You can use the **Note-Taking Checklist** before you go to lecture as a one-minute reminder of what to do to improve your note-taking skills. After lecture use it as a quality check.

After Lecture

The perfect time to review your notes is right after lecture. The following checklist contains both general review tips and specific suggestions for this chapter.

Check your notes: Have you...

☐ marked areas where you don't remember what was said? Do you need a follow-up session with your instructor, tutor, or study partner to get what you missed?

☐ added visual material? Example: For the lecture on Chapter 2, it is very important to distinguish the difference between diverging (Figure 2.6) and converging plates (Figure 2.9). You could insert a visual cue about this distinction into your notes. For example, the simplest possible representation of Figure 2.6 would be two arrows pointed away from each other. You also want to remember that divergence can cause both ridges and valleys. You could draw in ridges (zigzag lines) and valleys (dropping plates) between your arrows.

Study Tip: Learn by Drawing

Sketching simplified versions of figures into your notes is a helpful way to learn and remember. Visual learners will remember material best after they look at and study a figure. Visual learners learn more if they enrich their notes with visual cues. For kinesthetic learners memory is activated by the act of drawing. So you learn as you look and draw. As you take notes be sure to leave room to insert material later. The ultimate goal is to have a set of notes from which you can study.

There are two important caveats. First, you need to budget your time. Focus on very important material, such as the figures we suggested in the preview questions. You don't have time to redo all the artwork in *Understanding Earth*. Second, the art in the text is especially well done. It includes many details you cannot easily execute into your notes. Also, the captions in the text are very helpful. When you are reworking sketches be sure to refer to the apprpriate text figures and captions.

Get off to a good start. Try this idea for Chapter 2 and observe how it works for you. Modify it to fit your learning style.

Intensive Study Session

Set priorities for studying this chapter. There is a lot to do, quite likely more than you will have time for in one intensive study session. Set priorities and always do the important things first.

- **Instructor.** First, look to your instructor. Pay particular attention to any exercises recommended by your instructor during lecture and always answer those first. Your instructor is also your best resource if you are wondering which material is most important.
- **Practice Exercises and Review Questions.** Next, use the **Practice Exercises and Review Questions.** Be sure to do Exercise 1 because it gets to the key information you need to learn in this chapter.
- **Text.** Work on your responses to Exercises 3 and 5 and Thought Questions 1, 3, and 6 at the end of Chapter 12 in the textbook.
- **Web Site Study Resources**
  http://www.whfreeman.com/understandingearth

  Complete the **Concept Self-Checker** and **Web Review Questions.** Pay particular attention to the explanations for the answers. The **Geology in Practice** exercise, *New Neighbors?*, explores the movement of the San Andreas Fault and poses the question, "How long will it take before Los Angeles and San Francisco are neighbors?" **Flashcards** will help you learn the new terminology introduced in this chapter.

Exam Prep

Materials in this section are most useful during your preparation for midterm and final examinations.

The following **Chapter Summary** and **Practice Exercises and Review Questions** should simplify your chapter review. Read the **Chapter Summary** to begin your session. It provides a helpful overview that should refresh your memory.

Next, work on the **Practice Exercises and Review Questions.** In order to determine how you stand on mastery of this chapter, complete the exercises and questions just as you would a midterm. After you answer the questions score them. Finally, review any question that you missed. Identify and correct the misconception(s) that resulted in your answering the question incorrectly.

Chapter Summary

What is the theory of plate tectonics?

- For over the last century some geologists have argued for the concept of continental drift based on the jigsaw-puzzle fit of the coasts on both sides of the Atlantic, the geological similarities in rock ages and trends in geologic structures on opposite sides of the Atlantic, fossil evidence suggesting that continents were joined at one time, and the distribution of glacial deposits as well as other paleoclimatic evidence.

- In the last half of the Twentieth Century the major elements of the plate tectonics theory were formulated. Starting in the 1940s ocean floor mapping began to reveal major geologic features on the ocean floor. Then, the match between magnetic anomaly patterns on the seafloor with the paleomagnetic time scale revealed that the ocean floor had a young geologic age and was systematically older away from the oceanic ridge systems. The concepts for seafloor spreading, subduction, and transform faulting evolved out of these and other observations.
- According to the theory of plate tectonics, the Earth's lithosphere is broken into over a dozen moving plates. The plates slide over a hot and weak asthenosphere, and the continents, embedded in some of the moving plates, are carried along.

What are some of the geologic characteristics of plate boundaries?

- There are three major types of boundaries between lithospheric plates: divergent boundaries where plates move apart; convergent boundaries where plates move together and one plate often subducts beneath the other; and transform boundaries where plates slide past each other. Volcanoes, earthquakes, and mountains are concentrated along the active plate boundaries.

    Mountains typically form along convergent and transform plate boundaries. Where divergent plate boundaries are exposed on land, subsiding basins and volcanism are typical.

How can the age of the seafloor be determined?

- The age and relative plate velocity are inferred from changes in the paleomagnetic properties of the ocean floor.
- Various other methods are now used to measure the rate and direction of plate movements. Seafloor spreading rates vary between 2 and 24 cm per year today.

How is the history of plate movement reconstructed?

- Seafloor isochrons provide the basis for reconstructing plate motions for about the last 200 million years.
- Distinct assemblages of rocks characterize each type of plate boundary. Using diagnostic rock assemblages embedded in continents and paleoenvironmental data recorded by fossils and sedimentary rocks, geologists have been able to reconstruct ancient plate tectonics events and plate configurations.

What is the engine that drives plate tectonics?

- Driven by Earth's internal heat, convection within the mantle, coupled with the force of gravity, and the existence of a soft zone, called the asthenosphere, are important factors in models for how plate tectonics works.

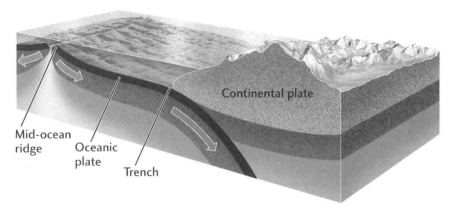

Figure 2.16. A schematic cross section through the outer part of the Earth, illustrating some of the forces thought to be important in driving plate tectonics: pulling force of sinking lithospheric slab; the pushing force of plates sliding off a mid-ocean ridge; the suction force of a retreating subduction zone.

Practice Exercises

Answers and explanations are provided at the end of the Study Guide.

Exercise 1: Characteristics of active tectonic plate boundaries.

Complete the table below by filling in the blank spaces and boxes.

Characteristics	Divergent See Figures 2.5, 2.6, 2.7, and 2.8.	Convergent See Figures 2.5 and 2.9.			Transform See Figures 2.5 and 2.10.
		Ocean/ Ocean	Ocean/ Continental	Collision	
Examples		Japanese Islands Marinas Trench Aleutian Trench		Himalayas and Tibetan Plateau	
Topography	Oceanic ridge, rift valley, ocean basins, ocean floor features offset by transforms, seamounts	Trench, island arc			Offset of creek beds and other topographic features that cross the fault
Volcanism		Present			Not characteristic

Exercise 2:

Construct a conceptual flowchart or diagram that illustrates modern ideas on how plate tectonics works. Include thorough captions or a written description of how it works.

Be sure the role played by the following important factors are addressed.

- Earth's internal heat
- convection
- asthenosphere
- divergent and convergent boundaries
- density differences
- lithosphere
- the push and pull of gravity

Hint: Refer to pages 44–47 in the text.

Review Questions

Answers and explanations are provided at the end of the Study Guide.

1. The most important process in building the ocean floor is
    A. volcanism.
    B. subduction.
    C. earthquake activity.
    D. magnetic reversal.

2. The youngest ocean crust is located
    A. along the oceanic ridges.
    B. in the oceanic trenches.
    C. around hot spots.
    D. on the abyssal seafloor.

3. Rates of seafloor spreading today are
    A. millimeters per year.
    B. centimeters per year.
    C. meters per year.
    D. kilometers per year.

    Hint: Refer to Figure 2.13.

4. Along a transform plate boundary the two plates
    A. move apart to create a widening rift valley.
    B. are being consumed by subduction.
    C. are being forced together so as to produce a mountain system.
    D. move horizontally past each other.

5. The _____ is an example of a divergent plate margin.
    A. East African Rift
    B. Japan Trench
    C. Himalayas
    D. San Andreas fault

6. All of the following features mark plate boundaries except the
    A. Red Sea.
    B. San Andreas Fault.
    C. Hawaiian Islands.
    D. Iceland.

    Hint: Refer to Figures 2.5 and 2.13.

7. Which of the following is not associated with a type of active plate boundary?
    A. Atlantic coast of North America
    B. Northwestern North America
    C. Gulf of California
    D. Himalayas

8. The name given to the supercontinent which broke up in the Early Jurassic (195 m.y.) was
    A. India.
    B. Atlantis.
    C. Laurasia.
    D. Pangaea.

9. Mid-ocean ridges, according to plate tectonics theory, are
    A. places where oceanic crust is consumed.
    B. pull-apart zones where new oceanic crust is produced.
    C. locations of plate convergence.
    D. transform faults.

10. The _____ are an example of a collision zone between two pieces of continental crust riding on converging lithospheric plates.
    A. Himalaya Mountains
    B. islands of Japan
    C. Aleutian Islands in Alaska
    D. Andes Mountains in South America

11. At which plate boundary are volcanoes least likely to form?
    A. Divergent boundaries
    B. Convergent boundaries
    C. Transform boundaries
    D. Hot spots

12. During the early Triassic (237 m.y.)
    A. India collided with Asia to form the Himalayan Mountains.
    B. the supercontinent, Rodinia, formed.
    C. the supercontinent, Pangaea, was mostly assembled.
    D. the Atlantic Ocean had already begun to open.
    Hint: Refer to Figure 2.15.

13. South America lay closest to the South Pole during
    A. Late Proterozoic (650 m.y.).
    B. the last 65 million years.
    C. Late Jurassic (152 m.y.).
    D. Early Devonian (390 m.y.).
    Hint: Refer to Figure 2.15.

14. The significance of the magnetic anomaly patterns discovered in association with the seafloor was that the anomaly patterns
    A. could be matched with the magnetic reversal chronology to establish an estimated age for the seafloor.
    B. provided evidence for mantle convection, a driving mechanism for plate tectonics.
    C. allowed geomagnetists to reconstruct the supercontinent Rodinia.
    D. represented absolute proof that the seafloor was spreading apart.

15. The oldest rocks on the seafloor are about
    A. 20 million years old.
    B. 200 million years old.
    C. 500 million years old.
    D. one billion years old.

16. The oceanic crust
    A. becomes progressively younger away from the oceanic ridges.
    B. becomes progressively older away from the oceanic ridges.
    C. is the same age virtually everywhere.
    D. ranges in age from Jurassic to Precambrian.

17. Volcanic island arcs, such as the Japanese Islands, are associated with
    A. convergent boundaries.
    B. divergent boundaries.
    C. transform boundaries.
    D. a chain of hot spots.
    Hint: Refer to Figure 2.9.

18. Rift valleys are associated with
    A. convergent boundaries.
    B. divergent boundaries.
    C. transform boundaries.
    D. active continental margins.
    Hint: Refer to Figures 2.6 and 2.9.

19. With what tectonic activity is a rift valley usually associated?
    A. Subduction.
    B. Movement on a transform fault.
    C. Continental rupture.
    D. Continental collision.

20. Magnetic anomalies in the seafloor are caused by
    A. magnetic reversals recorded by lavas erupted at oceanic spreading centers.
    B. changes in the atomic structure of minerals in response to changing ocean depth.
    C. the metamorphism of deep-sea sediments.
    D. the heating up of subducting oceanic lithosphere as it plunges deeper into the mantle.

21. Where do the plate-driving forces originate?
    A. Tectonic plates are passively dragged by convection currents.
    B. Gravity pulls and/or pushes old, cold, heavy lithosphere.
    C. Injection of magma from the mantle pushes the lithosphere apart.
    D. Earthquakes cause the plates to move.

22. Earth's lithosphere can be characterized as
    A. having an average thickness of about 100 km.
    B. including the crust and upper mantle.
    C. being solid and above the asthenosphere.
    D. all of the above.

CHAPTER 3

Minerals: Building Blocks of Rocks

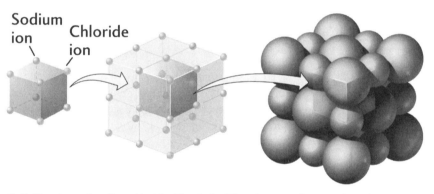

Figure 3.6. Structure of sodium chloride. The dashed lines between the ions show the cubic geometry of this mineral; they do not represent chemical bonds.

Before Lecture

Chapter Preview

- **What is a mineral?**
  Brief answer: A naturally occurring inorganic, crystalline solid.
- **How do atoms combine to form the crystal structure of minerals?**
  Hint: See structures for silicate minerals in Figure Story 3.11.
- **What is the atomic structure of minerals?**
  Brief answer: For silicate minerals it is various arrangements of silica tetrahedra.
- **What are the major rock-forming minerals and their physical properties?**
  Brief answer: The silicates, carbonates, oxides, sulfides, and sulfates. Diagnostic properties include hardness, cleavage, luster, color, density, and the shape of the crystal.

> **Previewing Tip**
>
> To preview a chapter quickly, focus on the figures. It is often possible to get a good overview just by examining the figures and captions. Note that preview questions often refer you to specific key figures. As little as 5 or 10 minutes previewing the figures before lecture will improve the quality of your note taking.

Vital Information from Other Chapters

- Review Figure 1.7 (The relative abundance of elements in the whole Earth with that of elements in Earth's crust) on page 8.

Minerals are the alphabet and rocks are the words.

—TOM DILLEY

During Lecture

Do a learning warm-up. Arrive early. Look for pictures in the text that catch your interest. Look over the preview questions and keep them handy for easy reference.

Your basic goal during lecture is to get a good set of notes. Those notes should contain in-depth answers to the preview questions. To avoid getting lost in details keep the "big picture" in mind. Chapter 3 explains how rocks are built out of minerals and how minerals are built out of atoms.

Your lecturer may show slides of mineral crystals and may pass around specimens of particular minerals. You will get more out of these if you sit close to the front row where you can see the sample rocks as they are discussed. Remember to focus carefully on clues the lecturer provides for recognizing a particular sample in the field. As you listen try to formulate at least one good question you can ask during the lecture.

After Lecture

Review Notes

Right after lecture, while the material is fresh in your mind, is the perfect time to review your notes. Review to be sure you understand all the key points and wrote them down in a form that will be readable later. One good check is that your notes should provide in-depth answers to all the **Chapter Preview** questions.

> **Study Tip: Intensive Study Session**
>
> The process of learning geology is similar to the procedure for building a house. Before each lecture construct a frame of questions. During lecture attach details and ideas to the frame. After lecture master those ideas during an intensive study session. By learning about Plate Tectonics you have laid a good foundation; but geology is a skyscraper with 23 floors (one for each chapter). Each chapter supports those above it. If you don't completely master this chapter the next one will be more difficult.
>
> Schedule about an hour after each lecture for intensive study. Devote this time to mastering key concepts. Mastery is not gained by just reading the text. Mastery occurs as the result of asking yourself questions and answering them. The **Practice Exercises**

and **Review Questions** are designed to help you reach the mastery level quickly. The greater the number of these exercises and questions you can work into your study schedule early in the course, the easier subsequent chapters will be.

Everything you do in life is worth infinite care and infinite effort.
—J. D. McDonald

Intensive Study Session

Here are a few suggestions for how you might do an intensive study session for this chapter.

- **Review Questions.** Start your study session by determining how much you already know. Try answering the **Review Questions.** Notice that the answers to these questions are at the end of the Study Guide, so you can check your answers. Often we provide some additional information along with the answer. The **Review Questions** are a great way to start studying because it will help you focus on what you need to work on. Afterwards you can go back and read the text concerning any points you missed.
- **Practice Exercises.** Immediately before the **Review Questions** you will find **Practice Exercises.** These exercises always focus on some key material that you will learn best by an interactive approach that requires you to think. For this particular chapter the exercises deal with understanding how the structure of a mineral determines its physical properties.
- **Text.** Work on Exercises 7 and 9 at the end of the chapter. Also work on Thought Questions 7, 8, and 9.
- **Web Site Study Resources**
  http://www.whfreeman.com/understandingearth
  A lot of the fun of geology has to do with figuring out the how's and why's behind interesting and beautiful natural phenomena. Go online and explore how the wondrous blue color of the Hope Diamond is related to its atomic structure and composition by doing the **Geology in Practice** exercises. Do the **Online Review Exercises**: Chemistry Review, which contains an interactive periodic table of the elements. Look at pictures of some of the minerals discussed in this chapter in the **Photo Gallery**. **Flashcards** will help you learn new terms. Complete the **Concept Self-Checker** and the **Web Review Questions** to assess your understanding of the chapter material.

Obviously you don't have time to do all these things for every chapter. The idea is to try out some of the tools and then decide which will be most helpful given your personal learning style.

"Crystal" comes from the Greek word Krustallos, which means ice. Is ice—water ice—a mineral?

Exam Prep

Materials in this section are most useful during your preparation for midterm and final examinations.

The following **Chapter Summary** and **Practice Exercises and Review Questions** should simplify your chapter review. Read the **Chapter Summary** to begin your session. It provides a helpful overview that should refresh your memory.

Next, work on the **Practice Exercises and Review Questions.** In order to determine how you stand on mastery of this chapter, complete the exercises and questions just as you would a midterm. After you answer the questions score them. Finally, review any question that you missed. Identify and correct the misconception(s) that resulted in your answering the question incorrectly.

Chapter Summary

What are minerals?

- Minerals are naturally-occurring inorganic solids with a specific crystal structure and chemical composition.

How do atoms combine to form the crystal structure of minerals?

- Minerals form when atoms or ions chemically bond and come together in an orderly, three-dimensional geometric array—a crystal structure. The character of the chemical bond can be ionic, covalent, or metallic. Ionic bonds are the dominant type of chemical bonds in mineral structures.

What is the atomic structure of minerals?

- The atomic structure is determined by how the atoms or ions pack together to form a crystalline solid. The packing of atoms depends on their size and the characteristics of chemical bonds between the atoms.

What are the major rock-forming minerals and their physical properties?

- The strength of the chemical bonds and the crystalline structure determines many of the physical properties, e.g., hardness and cleavage of minerals.
- Silicate minerals are the most abundant class of minerals in the Earth's crust and mantle.
    Isolated and chain-silicates—olivine, pyroxene, and hornblende
    Sheet silicates—micas and clay minerals
    Framework silicates—feldspar and quartz
- Other common mineral classes include carbonates, oxides, sulfates, sulfides, halides, and native metals.

Why study minerals?

- One can deduce the composition of rocks because minerals have a definite composition.
- One can infer the environment in which the rock formed because most minerals are stable only under certain conditions of temperature and pressure.
- For fun and profit. Gemstones are treasured for their beauty, color, and rarity. Many minerals are used in industrial processes and manufacturing.

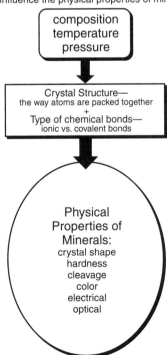

Conceptual flowchart illustrating the factors that influence the physical properties of minerals

SILICATE AND SILICATE POLYMORPH MINERALS

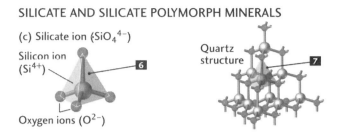

Figure 3.11. [PART C: Silicate ion]. A model showing the structure of the silicon-oxygen molecule. In the center of this model is one silicon ion carrying a positive charge of 4. It is surrounded by four oxygen ions, each carrying a minus charge of 2. After bonding, the resulting silicate ion carries a negative charge of 4 (4 minus 8 equals −4).

Practice Exercises and Review Questions

Answers and explanations are provided at the end of the Study Guide.

Practice Exercises

Exercise 1: Crystal Structures of Some Common Silicate Minerals

Three crystal structures of silicate minerals are illustrated next. The illustration provides a "birds-eye view" that looks down on a plane of atoms within the crystal structure of the mineral. The triangles, representing silica tetrahedra below, are equivalent to the blue pyramids in Figure Story 3.11.

- Fill-in-the-blank with the name for each major silicate structure (single chain, double chain, sheet, framework, or isolated tetrahedra).
- Give an example of a mineral with that crystal structure.
- Using the symbols given in the Cleavage Symbol Key below, characterize the cleavage properties for minerals with this crystal structure.

Hint: Refer to Figure Story 3.11 and page 59.

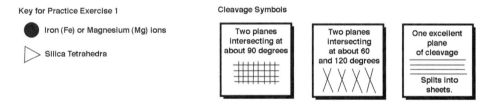

If you think the mineral will exhibit no cleavage, please write this inside the box.

A. Crystal structures of some common silicate minerals

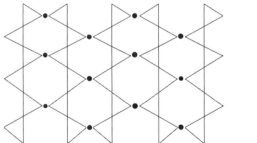

Basic Crystal Structure

Example Mineral

Cleavage (Use Cleavage Symbol Key as a guide for filling in the box.)

B. Crystal structures of some common silicate minerals

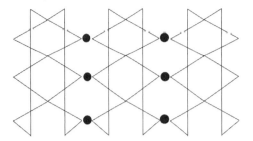

Basic Crystal Structure

_____

Example Mineral

_____

Cleavage (Use Cleavage Symbol Key as a guide for filling in the box.)

C. Crystal structures of some common silicate minerals

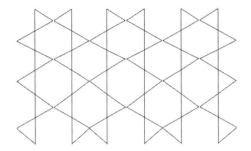

Basic Crystal Structure

_____

Example Mineral

_____

Cleavage (Use Cleavage Symbol Key as a guide for filling in the box.)

Exercise 2: Major Mineral Classes

Fill-in-the-blank with the name of the mineral class to which each mineral belongs.

The common classes of minerals include silicates, carbonates, sulfates, sulfides, oxides, native elements, and halides. **Hint:** Refer to Figure Story 3.11 and pages 60–62 in the textbook.

1. hornblende _silicate_
   Hint: an amphibole
2. spinel _____
3. clay minerals _____
4. calcite _____
5. gypsum _____
6. quartz _____
7. diamond _____
8. dolomite _____

13. graphite _native element_
14. pyrite _____
15. muscovite _____
    Hint: a mica
16. anhydrite _____
17. enstatite _____
    Hint: a pyroxene
18. pyrite _____
19. sapphire _____
20. orthoclase _____
    Hint: typically a tan-to-pink feldspar

9. plagioclase_____   21. silver _____
   Hint: a feldspar

10. ruby _____   22. albite _____
   Hint: typically a white feldspar

11. halite _____   23. olivine _____

12. hematite_____   24. biotite _____
   Hint: a black mica

A Little Geology Humor...

Why did the geologist skip the mineralogy luncheon?
   She didn't have any apatite.

The geology teacher asks a student if he can name the felsic mineral with crystal faces, conchoidal fracture, and a hardness of 7 on the Moh's scale.
   "Of quartz I can," says the student.

So, the geologist is at the eye doctor, getting fitted for a new pair of glasses. "Do you ever see double?" asks the doctor.
   "Only when I look through calcite samples in lab," says the geologist.

Exercise 3: Identifying Minerals Using Their Physical Properties

Give the mineral name for each of the mineral samples described below.
Hint: Refer to figures and text in Chapter 3, Appendix 3, and various links on minerals and gems provided on the web site: http://www.whfreeman.com/understandingearth.

Mineral Sample A
This mineral is colorless with a silvery shine. It separates into thin sheets or flakes that can bend without breaking. It is within a rock that contains large crystals of quartz and feldspar.
Hint: This mineral is named after Moscow where it was commonly used as a substitute for window glass.

Mineral name is _____

Mineral Sample B
This mineral sample is pale yellow and occurs as a vein within a rock fracture. Many small cubic crystals of this mineral were exposed in the vein when the rock broke apart along the fracture. The powdered form (streak) of the mineral is black and has a sulfur-like smell.

Hint: A common name for this mineral is "fool's gold."

Mineral name is _____

Mineral Sample C
This is on display at a mineral exhibition both as a mineral specimen and as cut and polished pieces in jewelry. It consists of beautiful concentric bands of various shades of green. A knife blade can easily powder its surface but my fingernail cannot scratch it. Acid will react with the mineral, especially if powdered. The specimen was labeled as coming from a copper mine in Bisbee, Arizona.

Hint: Refer to the web site's **Photo Gallery.**

Mineral name is _____

Mineral Sample D

This tan-to-pink sample occurs as many 5- to 7-centimeter crystals are surrounded by flakes of mica and grains of white-to-clear quartz. The tan-to-pink crystals are prismatic—rectangular/box shaped. It has a hardness between a knife blade and a steel file.

Hint: It is one of the most common minerals in the Earth's continental crust. A semiprecious variety, called moonstone, is used in jewelry.

Mineral name is _____

Mineral Sample E

Exhibiting excellent cleavage in three directions, this mineral breaks into beautiful rhombohedral-shaped pieces. Samples varied in color from clear to white to tan. A fingernail would not scratch the surface but a knife blade easily powdered the surface. Weak acid readily reacted with the surface of the mineral.

Hint: Marine organisms typically produce this mineral from seawater to generate their skeletons—seashells.

Mineral name is _____

Mineral Sample F

This sample was found as a series of thin layers with other layers of mud and silt. My fingernail could easily scratch its surface and form a white powder. Water dissolved the powdered form but the sample did not react with acid. Samples commonly exhibited a splintery aspect but would not separate into individual fibers, such as asbestos. The sample did easily break (cleave) along one plane but would not form the thin sheets typical of mica.

Hint: This mineral is a major constituent of plaster of Paris and wallboard used in buildings and houses.

Mineral name is _____

Review Questions

Answers and explanations are provided at the end of the Study Guide.

1. The bonds between Na and Cl in halite are strongly ionic. In ion form, Cl has seven electrons in its outer shell and Na has one. After these two elements bond, Cl has _____ electrons in its outer shell and Na has _____ electrons in its outer shell.

    A. six, two
    B. four, four
    C. one, seven
    D. eight, eight

    Hint: Refer to Figure 3.4.

2. _____ and _____ are examples of minerals with an identical chemical composition but different crystal structures.

    A. Calcite, dolomite
    B. Hematite, magnetite
    C. Pyrite, gypsum
    D. Graphite, diamond

3. The term geologists use for a naturally occurring aggregate of minerals is

    A. element.
    B. rock.
    C. compound.
    D. crystal.

4. When a substance is made of atoms that are arranged in a fixed, orderly and repeating pattern, it is said to be
   A. amorphous.
   B. glassy.
   C. crystalline.
   D. liquid.

5. Which of the following statements is NOT true of minerals?
   A. They are crystalline.
   B. They possess a definite chemical composition.
   C. They are naturally occurring.
   D. They are organic.

6. Which of the following objects is NOT a mineral?
   A. Native copper
   B. Glass
   C. Diamond
   D. Water ice

7. The most common mineral group on Earth is
   A. carbonates.
   B. silicates.
   C. oxides.
   D. halides.

8. The structure of feldspars, such as orthoclase and plagioclase, consists of
   A. double chains of silica tetrahedra.
   B. a three-dimensional framework of silica tetrahedra.
   C. single chains of silica tetrahedra.
   D. isolated silica tetrahedra.

9. Micas are known for their tendency to split apart into thin sheets; their silicate crystal structure is
   A. isolated silica tetrahedra.
   B. chains of silica tetrahedra.
   C. sheets of silica tetrahedra.
   D. a framework arrangement of silica tetrahedra.

10. Which of the following does *not* belong to the carbonate class of minerals?
    A. Calcite
    B. Anhydrite
    C. Dolomite
    D. Aragonite

11. Which of the following minerals has a sheet silicate structure?
    A. Clay minerals
    B. Feldspar
    C. Amphibole
    D. Graphite

12. Amphiboles and pyroxenes are mineral groups that belong to the class of minerals called
    A. chain silicates.
    B. framework silicates.
    C. sulfide minerals.
    D. sheet silicates.

13. Which of the following statements is *not* true about quartz and calcite?
    A. Calcite has excellent cleavage; quartz has poor to no cleavage.
    B. Quartz is a carbonate mineral.
    C. Both are often colorless.
    D. Quartz has a hardness of about 7 and calcite a hardness of about 3.

14. The characteristic of certain minerals to break along planes of weakness is called
    A. crystal symmetry.
    B. cleavage.
    C. hardness.
    D. luster.

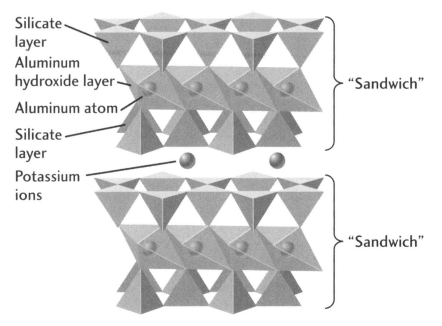

Figure 3.17. Cleavage of mica.

15. Graphite and diamond are two very different minerals. Why are the physical properties of graphite and diamond so different even though they are both made from pure carbon?
    A. The chemical bonds between the carbon atoms and their crystal structures are significantly different.
    B. Graphite has strikingly different physical properties from diamond because of impurities within its crystal structure, whereas diamond is pure carbon.
    C. Actually, graphite and diamond are not minerals because they are made from carbon and all matter made from carbon is organic. Their physical properties are different because they were produced by very different living organisms.
    D. None of the above.

    Hint: Refer to Figure Story 3.11 (a and b).

16. Chemical bonding along cleavage planes within the crystal structure of a mineral typically
    A. is more covalent.
    B. is more ionic.
    C. is more magnetic.
    D. involves electron sharing between atoms.

    Hint: Refer to the section in the textbook titled *Cleavage*.

17. Iron and magnesium ions are similar in size and both have a +2 charge. Therefore, we would expect
    A. iron and magnesium to bond easily.
    B. iron and magnesium to share electrons.
    C. iron and magnesium to form minerals with different crystal structures.
    D. iron and magnesium to substitute for each other within the crystalline structure of minerals.

    Hint: Refer to Figure 3.7 and the section in the textbook titled *How Do Minerals Form*.

18. The chemical bonds between carbon atoms within diamond are predominantly
    A. covalent.
    B. ionic.
    C. metallic.
    D. nuclear.

CHAPTER 4

Rocks: Records of Geologic Processes

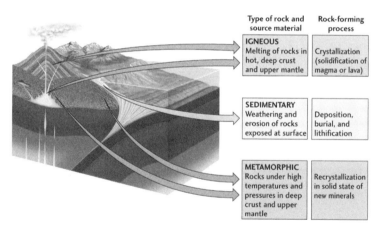

Figure 4.2. The three different types of rock.

Before Lecture

Prelecture Chapter Preview

Before you attend the lecture be sure to spend some time previewing Chapter 4. Use the following Chapter Preview questions as a guide. These questions constitute the basic framework for understanding this chapter. Working with the preview questions before lecture and committing them to memory should help you to understand the lecture much better and get an excellent set of notes. Need a refresher on previewing? See the flowchart **"How to Study Geology"** in Chapter 1 (Part I) of this Study Guide. This chapter integrates all study strategy suggestions from this guide into one essay.

Chapter Preview

- **What determines the properties of rocks?**
  Brief answer: Mineral content and texture.
- **What are the three types of rocks and how are they formed?**
  Brief answer: Igneous, sedimentary, and metamorphic rock.
- **What is the association of different rock types with plate tectonics?**
  Brief answer: Refer to Figures 4.2 and 4.6 and Figure Story 4.9.
- **How does the rock cycle describe rock formation and the relationship between different rock types?**
  Hint: Study Figure Story 4.9.

Vital Information from Other Chapters

Plate tectonics processes (Chapter 2) drive the formation of igneous and metamorphic rocks in addition to the burial sediments. It will be particularly useful to review Chapter 2, especially text and figures regarding converging and diverging plate boundaries.

During Lecture

- Your basic goal during lecture is to get a good set of notes. By the end of the lecture those notes should contain in-depth answers to the **Chapter Preview** questions. It may be helpful to keep a copy of the preview questions in front of you (better yet, develop preliminary answers to the questions and commit them to memory before the lecture).
- To avoid getting lost in details keep the "big picture" in mind. Chapter 4 tells the story of the three basic types of rock and how these are formed by the geologic processes of the rock cycle.
- Want ideas on taking a good set of notes? If you haven't already done so, read our discussion of note taking in **How to Be Successful in Geology** in Chapter 3 (Part I) of this Study Guide. **Hint**: You can use the **Note-Taking Checklist** before you go to lecture as a 1-minute reminder of what to do to improve your note-taking skills. After lecture use it as a quality check.

Lecture Tip: Rock Texture

Your instructor may show or pass around samples of the three types of rocks to help you learn how they differ. Pay particular attention to the crystal or grain size (texture), presence of layering, fossils, and patterns produced by the alignment of minerals. You will have a better chance of seeing the samples if you sit at the front of the room.

The important thing is not to stop questioning.

—ALBERT EINSTEIN

After Lecture

The perfect time to review your notes is right after lecture. The following checklist contains both general review tips and specific suggestions for this chapter.

Check your notes: Have you...

☐ added a comparison chart to help you sort out the key differences between the three rock types? (See example below.)

☐ added simplified sketches of key text figures to your notes? It turns out that trying to simplify a figure is often the best way of studying it. It gets you involved in the important details. Example: Draw a simplified version of Figure 4.3 to help you remember the difference in texture (crystal size) between intrusive and extrusive igneous rocks.

Major rock type	Igneous	Sedimentary	Metamorphic
Source of materials	Melting of rocks in crust and mantle.	Weathering and erosion of rocks exposed at the surface.	Rocks under high temperatures and pressures in deep crust and upper mantle. Plus rocks near the surface that are broken by faulting or impact.
Rock-forming process	Crystallization (solidification of melt).	Sedimentation, burial, and lithification (compaction and cementation).	Shearing and recrystallization in the solid state.

Refer to Figure 4.2 (The three different types of rocks).

Intensive Study Session

Set specific priorities for studying this chapter. We recommend you give highest priority to activities that involve answering questions. Answering questions while using your text and lecture notes as reference material is far more efficient than reading chapters or glancing over notes.

- **Practice Exercises and Review Questions.** Begin with the **Practice Exercises and Review Questions**. Be sure to do Exercise 1. It focuses on key information you need to learn in this chapter.
- **Web Site Resources**
  http://www.whfreeman.com/understandingearth

  Complete the **Concept Self-Checker** on the web site. Pay particular attention to the explanations for the answers. Also on the web site are **Flashcards** to help you learn new terms. You may want to check out this feature for future reference. The **Online Review Exercise** *Rock Cycle Review* provides an excellent tour through the rock cycle and how it works in the context of plate tectonics. Use it for review. Practice distinguishing the diagnostic characteristics of the basic rock types by completing the **Geology in Practice** exercises.

Exam Prep

Materials in this section are most useful during your preparation for midterm and final examinations.

The following **Chapter Summary**, and **Practice Exercises and Review Questions** should simplify your chapter review. Read the **Chapter Summary** to begin your session. It provides a helpful overview that should refresh your memory.

Next, work on the **Practice Exercises and Review Questions.** In order to determine how you stand on mastery of this chapter, complete the exercises and questions just as you would a midterm. After you answer the questions score them. Finally, review any question that you missed. Identify and correct the misconception(s) that resulted in your answering the question incorrectly.

Chapter Summary

- **What determines the properties of rocks?**
  The properties of rocks and the rock's name are determined by mineral content (the kinds and proportions of minerals that make up the rock) and texture (the size, shapes, and spatial arrangement of its crystals or grains).
- **What are the three types of rocks and how are they formed?** There are three major rock types. Igneous rocks solidify from molten liquid (magma). Crystal size within igneous rocks is largely determined by the cooling rate of the magma body. Sedimentary rocks are made of sediments formed from the weathering and erosion of any preexisting rock. Deposition, burial, and lithification (compaction and cementation) transform loose sediments into sedimentary rocks. Metamorphic rocks are produced by alteration in the solid state of any preexisting rock by high pressures and temperatures, which result in a change in texture, mineral composition, or chemical composition.
- **How does the rock cycle describe rock formation and its relationship to plate tectonics?** The Rock Cycle is the result of interactions between plate tectonics and climate.

Refer to Figure Story 4.9.

Practice Exercises and Review Questions

Answers and explanations are provided at the end of the Study Guide.

Exercise 1: Rock cycle review

Information and figures in Chapters 2 and 4 will be a helpful reference for this exercise.

A. What are the three plate tectonic settings for the generation of magma?

   hot spots/mantle plumes, such as Hawaii

   _____

   _____

B. Cooling rates and textures of igneous rocks

   Given the two types of igneous rocks (extrusive and intrusive) complete the table with their characteristic cooling rates (*fast vs. slow*) and textures (*fine-grained vs. coarse-grained*).

Types of igneous rocks	Cooling rates	Textures
Extrusive		
Intrusive		

C. There are three major agents for the transport and deposition of sediments on land. They are:

ice

_____

_____

D. What are the two ways loose sediments are transformed into solid sedimentary rock?

_____

_____

E. What are the two main types of sedimentary rocks and what are they made out of?

Sedimentary rock type	What are they made out of?

F. What are the four major conditions (geologic settings) that result in metamorphic rocks?

_____

_____

_____

Hint: Refer to Figure 4.6.

G. Are metamorphic rocks formed by melting? Explain.

_____

_____

_____

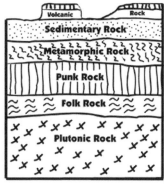

Study Tip: Construct a Simplified Rock Cycle

Figure Story 4.9 packs much information into one page and provides an overview of how rock types are linked to Earth's plate tectonic and climate system. As a review for an exam, construct your own rock cycle flowchart. Keep it relatively simple and focus on presenting the following information:

- Major rock types
- Highlight one or two important diagnostic characteristics of each rock type
- Process(es) that transform a rock into another rock type

Use sketches, words, phrases, or bulleted lists to describe each component (each major rock type) and connecting pathways (processes that transform rocks into another type).

If you are a visual learner, concentrate on drawing simple illustrations that characterize how each rock type forms and is linked to other rock types. If you learn better from reading or hearing, then use arrows and describe the major elements of the rock cycle in words and read it out loud.

Review Questions

Answers and explanations are provided at the end of the Study Guide.

1. Which of the following rocks form from molten material cooling and solidifying within the Earth's crust?
    A. Volcanic
    B. Plutonic
    C. Sedimentary
    D. Metamorphic

2. Molten rock within the Earth's crust is called
    A. silica.
    B. lava.
    C. magma.
    D. mica.

3. Coarse-grained igneous rocks, such as granite, are exposed today at the Earth's surface due to
    A. uplift and erosion.
    B. quickly cooled lavas erupting from ancient volcanoes.
    C. silicate minerals precipitated from rainwater.
    D. all of the above.

4. Lithification is the process that converts sediments into solid rock by
    A. cooling and crystallization.
    B. pressure cooking.
    C. subduction.
    D. cementation and compaction.

5. The texture of igneous rocks reflects the
    A. rate of cooling.
    B. tectonic forces that have altered the minerals.
    C. rate of sedimentation.
    D. composition.

6. Bedding (layering) is a major identifying characteristic of
    A. sedimentary rocks.
    B. metamorphic rocks.
    C. intrusive rocks.
    D. none of the above.

7. The rock type that covers most of the Earth's land surface and seafloor is
    A. extrusive.
    B. intrusive.
    C. metamorphic.
    D. sedimentary.
    Hint: Refer to Figure 4.5.

8. Contact metamorphism
    A. occurs in areas of very high temperature and pressure, like in subduction zones.
    B. extends over very large areas.
    C. is characteristic of continental collision zones.
    D. occurs in areas where magma intrudes and metamorphoses neighboring rock.
    Hint: Refer to Figure 4.6

9. In the rock cycle weathering
    A. creates sediment.
    B. results in burial and lithification.
    C. creates mountains.
    D. can cause metamorphism.
    Hint: Refer to Figure Story 4.9.

10. In the rock cycle mountains typically form as a result of
    A. weathering and erosion.
    B. sedimentation piles up rock.
    C. subduction.
    D. metamorphism.

11. It is reasonable to assume that there are no sedimentary rocks on our moon because it
    A. has no surface water and atmosphere.
    B. experienced no volcanism.
    C. lacks an active tectonic system.
    D. lacks any evidence for metamorphism.

CHAPTER 5

Igneous Rocks: Solids from Melts

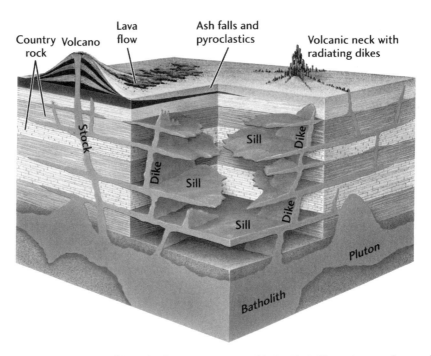

Figure 5.7. Basic extrusive and intrusive igneous structures. Notice that dikes cut across layers of country rock, but sills run parallel to them. Batholiths are the largest form of plutons.

Before Lecture

Before you attend the lecture be sure to spend some time previewing Chapter 5. We have made it easier by identifying the five key questions on igneous rocks for you (see **Chapter Preview** questions). These questions constitute the basic framework for understanding this chapter. Working with the **Preview** questions before lecture and committing them to memory should help you to understand the lecture much better and get an excellent set of notes. Need a refresher on previewing? See the flowchart **How to Study Geology** in Chapter 1 (Part I) of this Sudy Guide.

Chapter Preview

- **How are igneous rocks classified?**
  Brief answer: Composition and texture (cooling history). Preview Figures 5.1, 5.3, and 5.4 and Tables 5.1 and 5.2. **Hint:** Table 5.2 summarizes the crucial information, but you will need to look at the other figures in order to understand it.
- **How and where do magmas form?**
  Hint: Table 5.3 and Figure 5.11 in the textbook provide an overview.
- **How does magmatic differentiation account for the great variety of igneous rocks?**
  Hint: Figure Story 5.5 tells the classic differentiation story. Figure 5.6 adds important details.
- **What are the forms of intrusive and extrusive igneous rocks?**
  Hint: Figures 5.7 and 5.10 show the basic intrusive and extrusive igneous rock bodies.
- **How do igneous rocks relate to plate tectonics?**
  Brief answer: Magmas tend to form at divergent and convergent plate boundaries and hot spots. **Hint:** Refer to Figures 5.11, 5.13, and 5.14 and the "Activity, Rock Composition, and Types of Magmas" section of Chapter 5 for more details.

Plate tectonic setting (Examples)	Magma type (example rock type)
Spreading centers	
• oceanic ridge (Mid-Atlantic Ridge)	• mafic (basalt)
• continental rift (East African Rift)	• mafic to felsic (silicic)—more variable because some continental crust may melt.
Subduction zones	
• oceanic island arc (Japan)	• mafic to intermediate
• continental volcanic arc (Cascade Range, Mt. St. Helens, Mount Rainier, and Andes in South America)	• mafic to felsic (silicic)—more variable because continental crust may melt.
Note: Not much volcanism occurs in association with collision plate boundaries.	
Intraplate mantle plumes ("hot spots")	
• oceanic hot spots (Hawaii)	• mafic (basalt)
• continental hot spots (Yellowstone)	• mafic to felsic (silicic)—more variable because continental crust is melted.

Vital Information from Other Chapters

Review Figure 4.3 and Figure Story 4.9.

On the Web Site

http://www.whfreeman.com/understandingearth

Review the **Online Exercise:** *Rock Cycle Review* from Chapter 4 on the web site. Pay particular attention to information on melting, magmas, cooling, and igneous rocks.

During Lecture

A lot of important visual material will be covered in this lecture. Be sure you preview the chapter before the lecture. Because the material is somewhat technical you will want to approach taking notes in as organized a manner as possible. Following are some ideas you may find helpful.

- **Big picture for Chapter 5.** To avoid getting lost in details keep the "big picture" in mind throughout the lecture: Chapter 5 explains how plate tectonics drives the formation of magma and how igneous rocks of varying composition and texture are produced at particular plate locations. For example, fine textured volcanic rock of basaltic composition is produced at a diverging ocean plate (see Figures 5.13 and 5.14).
- **Key figures for lecture.** There are many new words and concepts to learn for this chapter. It may be helpful to put bookmarks in your text for one or two of the most important figures you previewed before coming to lecture. Figure 5.3 will help you with igneous rock classification. Figures 5.11, 5.13, and 5.14 summarize where and how magmas are formed.
- **Develop a new skill.** You will be learning how to identify igneous rocks. Often instructors use slides or rock samples to help you learn this skill. The following tip and chart tool will help you master this skill efficiently.

Note-Taking Tip: How to Take Notes on Rock Samples

During lecture your instructor may show slides of igneous rock formations and may pass around specimens of igneous rocks. These will be easier to see if you sit close to the front of the room. For your instructor each sample is similar to a chapter in a book. It tells a story. You can become proficient at reading rock stories too. It just takes practice and a little organization. For igneous rocks the following chart will help you to organize what you see and hear. It will help you to focus on the two things you need to give the most attention—*texture* (whether or not you can see crystals) and *composition* (what is the likely mineral content of this rock?). If your instructor tells an interesting story about the rock, make a note or two; stories help us remember details we might otherwise forget.

   What to do. Before the lecture on igneous rocks make a chart similar to the following one. Use a full sheet of notebook paper and leave plenty of room for notes about each sample. Try to get something in every column for every rock sample. If you miss a column or were unable to see some of the minerals in the rock, be sure to talk to your instructor right after lecture (while your memory is fresh and the rock sample is handy) and fill in what you need to complete your chart. **Hint:** This idea is going to be useful in subsequent chapters on sedimentary and metamorphic rocks.

Example: Rock Sample Note-Taking Chart

(Set up on one full page of your notebook)

Rock name	Texture clues (coarse, fine, etc.)	Composition clues (mineral clues, color, dark/light, etc.)	Story about the rock to help you remember it	Other notes
Example: **Granite**	Coarse	I could see crystals of black hornblende, white and pink feldspars, and thin, shiny sheets of mica in a light gray matrix (quartz).	"Came from a local (you name it) mountain near our campus. I slipped and fell just before I found this sample." Stories will help you to remember information.	Crystals were clearly visible in the slide. Crystals were mostly feldspar and muscovite mica. What I thought was hornblende turned out to be biotite. I tried to scratch the rock with a penny, no luck, must be kind of hard (above 3 on the Mohs scale). Minerals in granite average greater than 5 on the Mohs Scale. Too bad there wasn't a bowie knife in the classroom.

Sample # 1

·
·
·
·
·
·

Sample #?

After Lecture

Because Chapter 5 gets a bit technical it will be particularly important for you to rework and improve your notes after lecture.

> **Check your notes: Have you...**
>
> ☐ added the five preview questions as section headings for your notes? You should have sections dealing with 1) classification, 2) magma formation, 3) magma differentiation, 4) igneous structures, and 5) plate tectonics and igneous rock formation.

☐ added exercise material from this guide to your notes? A great place to begin is by completing Exercise 1 (see **Practice Exercises and Review Questions**). Exercise 1 will help you understand what is perhaps the single most fundamental idea about igneous rock, namely rock texture. If you do the exercise sketches on notebook paper, you can clip them in somewhere near the beginning of your notes on igneous rocks for future review.

☐ added visual material to your notes? Key figures to consider adding are Figure 5.4: Classification model of igneous rocks; Figure Story 5.5: Magma; Figure 5.7: Basic extrusive and intrusive igneous structures; and Figure 5.11: Plate tectonics and magma formation. Remember, the idea is to draw a simplified version that emphasizes the features discussed in lecture.

☐ included a brief summary of the big picture for this chapter? Writing a brief summary of the essence of this chapter is a good way to help focus on what is important and avoid getting bogged down in less essential details.

Intensive Study Session

Set specific priorities for studying this chapter. We recommend you give highest priority to activities that involve answering questions. Answering questions while using your text and lecture notes as reference material is far more efficient than reading chapters or glancing over notes.

- **Text.** There are some great features built into this chapter to be used for learning purposes. One good example is Figure 5.2. Read the section on *Texture* at the beginning of the chapter. You will learn how James Hutton deduced the nature of igneous rock by assessing three lines of evidence (clues 1–3). Then, look at Figure 5.2. You can have an experience of discovery not unlike Hutton's. Study Figures 5.1, 5.2, and 5.3 until you think you understand texture. Then, do Exercise 1 to master how to interpret rock textures. Move on through the chapter, focusing particular attention on Figures 5.3, 5.4, 5.5, 5.7, and 5.11.

- **Practice Exercises and Review Questions.** Next, use the **Practice Exercises and Review Questions.** Be sure to do Exercise 1. It focuses on the key information you need to learn in this chapter.

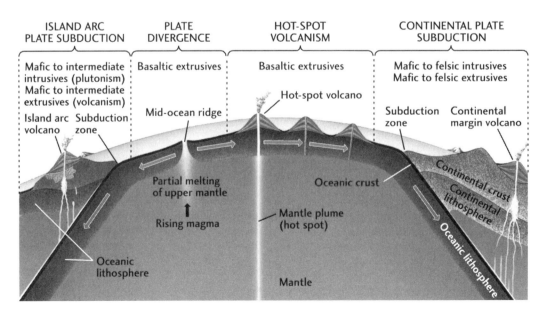

Figure 5.11. Magma forms under conditions that are strongly connected to movements of lithospheric plates. These movements control where rocks of the crust and upper mantle melt and whether they will be intruded or extruded.

- **Web Site Resources**
  http://www.whfreeman.com/understandingearth
  Complete the following **Online Review Exercises**: *Bowen's Reaction Series Review* and *Igneous Rock Review*. **Flash Cards** will help you learn new terms. Try the **Concept Self-Checker.** Also, experiment with being a field geologist by doing the **Geology in Practice Exercises,** which will provide an excellent review of basic information on igneous rocks. The photomicrographs in the **Photo Gallery** will be very helpful in both understanding the differences in grain size for many rocks and for completing Exercise 1.

Exam Prep

Materials in this section are most useful during your preparation for midterm and final examinations.

The following **Chapter Summary and Practice Exercises and Review Questions** should simplify your chapter review. Read the **Chapter Summary** to begin your session. It provides a helpful overview that should refresh your memory.

Next, work on the **Practice Exercises and Review Questions.** In order to determine how you stand on mastery of this chapter, complete the exercises and questions just as you would a midterm. After you answer the questions score them. Finally, review any question that you missed. Identify and correct the misconception(s) that resulted in your answering the question incorrectly.

Chapter Summary

How are igneous rocks classified?

- Igneous rocks can be divided into two broad textual classes: 1) coarsely crystalline rocks, which are intrusive (plutonic) and, therefore, cool slowly, and 2) finely crystalline rocks, which are extrusive (volcanic) and cool rapidly. Within each of these broad textual classes, the rocks are subdivided according to their composition. General compositional classes of igneous rocks are felsic, intermediate, mafic, and ultramafic, with decreasing silica and increasing iron and magnesium content.

How and where do magmas form?

- The lower crust and upper mantle are typical places where physical conditions induce rock to melt. High temperatures, a reduction in pressure, and the presence of water all cause melting.
  Composition is also a factor in the melting temperature of rocks.

How to melt a rock—the generation of magma

- Increase temperature
  Not all minerals melt at the same temperature (refer to Figure Story 5.5). The mineral composition of the rock affects the melting temperature. Felsic rocks with higher silica content melt at lower temperatures than mafic rocks with less silica and more iron and magnesium.
- Lower the confining pressure
  A reduction in pressure can induce a hot rock to melt. A reduction in confining pressure on the hot upper mantle is thought to generate the basaltic magmas; these intrude into the oceanic ridge system to form ocean crust (refer to Figure 5.13).

- Add water

    The presence of water in a rock can lower its melting temperatures up to a few hundred degrees.

    Water released from rocks subducting into the mantle along convergent plate boundaries may be an important factor in magma generation, especially at convergent plate boundaries (refer to Figure 5.14).

How does magmatic differentiation account for the great variety of igneous rocks?

- There is an amazing variety of igneous rocks on Earth. Two processes help to explain how the composition of igneous rocks can be so variable. These are partial melting and fractional crystallization (refer to Figure Story 5.5 and Figure 5.6).
- The Bowen's Reaction Series in Figure Story 5.5 is a flowchart describing how the very general bulk composition of a magma can change as the magma solidifies or as a rock melts.

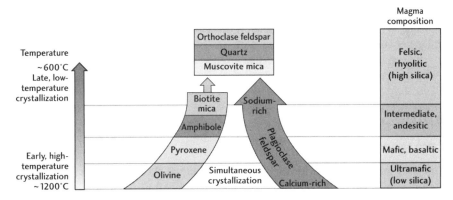

Figure Story 5.5. Bowen's Reaction Series provides a model for the sequence in which magmas crystallize and rocks melt. Fractional crystallization and partial melting result in magmas with a different composition than the parent rock from which the melt was derived.

What are the forms of intrusive and extrusive igneous rocks?

- Names are given to igneous rock bodies based on their size and shape. Figure 5.7 summarizes the common igneous rock bodies, such as batholith, pluton, dike, and sill.

How do igneous rocks relate to plate tectonics?

- Mid-ocean ridges and subduction zones are the major sites of magmatic activity (refer to Figures 5.11, 5.13, and 5.14).

Practice Exercises and Review Questions

Answers and explanations are provided at the end of the Study Guide.

Exercise 1: Igneous rock textures

In the four boxes, sketch the igneous rock texture that is consistent with the origin of the rock as described next to each box. Fill in each blank with the appropriate texture term from the list below. Table 5.2, Figure 5.3, and Figure Story 4.9 will be helpful.

Texture terms

Fine grained (aphanitic)

Intermediate grain sizes—visual grains but not very coarse grained

Coarse grained (phaneritic)

Mixture of coarse and fine grains (porphyritic)

A. Draw the texture of an igneous rock from a pluton solidified at depth within the crust.

phaneritic
name of texture

B. Draw the texture of an igneous rock from a shallow magma body, such as a dike or sill.

name of texture

C. Draw the texture of a lava erupted from a magma chamber after some minerals had begun to crystallize

name of texture

D. Draw the texture of a lava flow that erupted before the magma chamber underwent any crystallization.

name of texture

Exercise 2: Distribution of igneous rocks within the Earth

Complete this table by filling in the blanks using terms from the lists below. Refer to Table 5.2 and Figure 5.4 in your textbook. Some answers are provided. Note that the Earth's core is not included in this table because it is thought to be composed mostly of iron and nickel and not silicate minerals.

Major layer within the Earth	Example of an igneous rock	General compositional group	General chemical composition
Continental crust (For continental crust, there are two appropriate answers.)	*Granite*		
		Intermediate	
Ocean crust			
Mantle			*Less Si, Na, K* *More Fe, Mg, Ca*

Example igneous rocks
granite/rhyolite
basalt/gabbro
andesite/diorite
peridotite
granodiorite/dacit

General compositional group
ultramafic
felsic
intermediate
mafic

General chemical compositions of igneous rocks
More (higher amounts) Si, Na, K
More (higher amounts) Fe, Mg, Ca
Less (lower amounts) Si, Na, K
Less (lower amounts) Fe, Mg, Ca
Intermediate

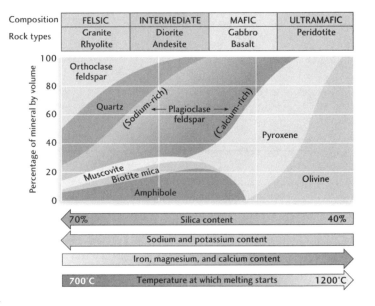

Figure 5.4. Classification model of igneous rocks. The vertical axis measures the mineral composition of a given rock as a percentage of its volume. The horizontal axis is a scale of silica content by weight. Thus, if you know by chemical analysis that a coarsely textured rock sample is about 70 percent silica, you could determine that its composition is about 6 percent amphibole, 3 percent biotite, 5 percent muscovite, 14 percent plagioclase feldspar, 22 percent quarz, and 50 percent orthoclase feldspar. Your rock would be a granite.

Exercise 3: Predicting the Change in Composition in a Crystallizing Magma

A basaltic magma intruded between sandstone layers to form the sill. As the magma cooled in place, fractions of the crystals that formed settled into layers—it underwent fractional crystallization. Using Figure 5.4, Table 5.2, and Figure Story 5.5, predict how the silica and iron content of the Palisades sill changes. Indicate whether the silica and iron content increase, decrease, or stay the same in each layer within the sill. Explain the reason(s) for your answers.

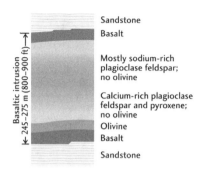

Circle the correct answer.
Olivine → calcium-rich plagioclase feldspar and pyroxene with no olivine

- A. Silica content: increased/decreased
- B. Iron content: increased/decreased

Explanation:

Calcium-rich plagioclase → sodium-rich plagioclase feldspar and no olivine

- C. Silica content: increased/decreased
- D. Iron content: increased/decreased

Explanation:

Exercise 4: Minerals and magma solidification

Circle the answers that correctly complete the following statements.

- A. The atomic (crystal) structure of the earliest formed silicate minerals in a magma tends to be MORE/LESS complex than the crystalline structures of minerals formed during later stages in the solidification of the magma.
- B. During the solidification of a magma, the minerals with the highest silica content will crystallize FIRST/LAST.
- C. In the last stages of solidification of a magma, the remaining silicate melt will contain MORE/LESS silica than the original melt.

    Hint: You do not need to know the actual bulk composition (mafic, intermediate, felsic) of the magma to answer these questions. Refer to Figure Stories 3.11 and 5.5.

Exercise 5: Partial melting and magma composition

Circle the answers that correctly complete the following statements.
   Compared to the bulk composition of the rock, the minerals with the lower melting temperature are:

- A. HIGHER/LOWER in the Bowen's Reaction Series.
- B. DEPLETED/ENRICHED in Mg, Fe, and Ca.
- C. DEPLETED/ENRICHED in Si, Na, and K.

    Hint: Refer to Figure Story 5.5.

Exercise 6: Predicting the composition of magma generated in subduction zones

Circle the answers (MORE/LESS) for each statement (A–D) that correctly completes the following sentence.

Compared to the basaltic ocean crust, the magma generated by partial melting of the subducting slab of ocean crust will:

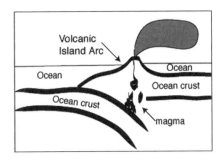

A. have MORE/LESS silica.
B. have MORE/LESS iron and magnesium.
C. have MORE/LESS sodium and potassium.
D. be MORE/LESS mafic.

Review Questions

Answers and explanations are provided at the end of the Study Guide.

1. Igneous rock names are based on
   A. texture and composition.
   B. fine-grained and coarse-grained.
   C. intrusive and extrusive.
   D. where the magma chamber erupts.

2. What rock has the same mineralogy as granite, but has a fine-grained texture?
   A. Andesite   C. Obsidian
   B. Basalt   D. Rhyolite
   Hint: Refer to Table 5.2.

3. Which of the following rocks contains the most silica?
   A. Basalt   C. Fissure eruptions
   B. Rhyolite   D. Dacite

4. Which of the following pairs of intrusive and extrusive rocks are made from the same minerals (i.e., they have the same chemical composition)?
   A. Gabbro and basalt   C. Granite and andesite
   B. Diorite and basalt   D. Gabbro and rhyolite

5. In the field or in hand-sized specimens intrusive and extrusive igneous rocks are distinguished by which characteristic?
   A. Composition   C. Porphyritic versus non-porphyritic texture
   B. Color   D. Grain size

6. Which of the following iron-rich mineral is most common in basalt?
   A. Pyroxene   C. Muscovite
   B. Quartz   D. Na-plagioclase

7. Granite is mainly made up of
   A. quartz, orthoclase (K-feldspar), and Na-plagioclase.
   B. quartz, Ca-plagioclase, Na-plagioclase, and amphibole.
   C. quartz, pyroxene, and muscovite.
   D. quartz, orthoclase (K-feldspar), Ca-plagioclase, and olivine.
   Hint: Refer to Table 5.1 and Figure 5.4.

8. An igneous rock made of a mixture of both coarse and fine grain minerals is called porphyritic and is formed by
   A. rapid cooling followed by a period of slow cooling.
   B. slow cooling followed by a period of rapid cooling.
   C. very slow cooling of a water-rich magma.
   D. very rapid cooling in the presence of water.

9. Only iron and magnesium-rich minerals are found in which of the following lists of minerals?
   A. Pyroxene, hornblende, K-feldspar, biotite
   B. Plagioclase, biotite, pyroxene, clay
   C. Quartz, muscovite, biotite, plagioclase
   D. Biotite, pyroxene, olivine, hornblende

10. Following Bowen's Reaction Series, the later lower-temperature fractions of liquid magma become progressively
    A. depleted in silica.
    B. enriched in silica.
    C. enriched in magnesium.
    D. depleted in potassium.
    Hint: Refer to Figure Story 5.5.

11. Which of the following minerals are the earliest highest-temperature minerals to crystallize in Bowen's Reaction Series?
    A. Quartz and feldspar
    B. Plagioclase and amphibole
    C. Plagioclase and olivine
    D. Chert and mica

12. The formation of granitic batholiths occurs
    A. within the ocean crust.
    B. within the continental crust.
    C. along spreading centers in the ocean.
    D. under hot spots.

13. Considering the following minerals, which pairs would you predict NOT to be found together in the same igneous rock?
    A. K-feldspar and biotite
    B. Na-plagioclase and muscovite
    C. Quartz and Na-plagioclase
    D. Quartz and olivine
    Hint: Refer to Figure 5.4 and Figure Story 5.5.

14. How would you distinguish a lava flow from a sill exposed at the Earth's surface?
    A. Sills tend to be coarser-grained than lava flows because they cool slower.
    B. Their chemical compositions (i.e., the minerals present) would be very different.
    C. Sills tend to be finer-grained due to slower rates of crystallization.
    D. Lava flows are coarser-grained due to very rapid rates of cooling.

15. The rock type of most batholiths found in the continental crust is
    A. gabbro.
    B. obsidian.
    C. granite.
    D. basalt.

16. How does a rising magma make space for itself as it moves through the solid crust?
    A. By breaking off large blocks of rock that sink into the magma chamber
    B. By wedging open the overlying rock
    C. By melting surrounding rocks
    D. All of these

17. The source for most mafic magmas is thought to be
    A. partial melting of felsic and intermediate rocks in the upper continental crust.
    B. partial melting of ultramafic rocks within the upper mantle.
    C. melting of preexisting granites and sediments.
    D. directly from the molten core of the Earth.

18. If lava flows on the slopes of a volcano are derived from one large magma chamber, which crystallizes slowly and feeds eruptions over a period of many thousands of years, how would you predict the gross composition of the lava to change as the lava flows become younger?
    A. Younger lava flows would become progressively enriched in iron and become more mafic.
    B. Younger lava flows would become progressively enriched in silica and become more felsic.
    C. Lava flows would be the same composition because they all came from the same magma chamber.
    D. Lava flows would alternate in composition.
    Hint: Use the Bowen's Reaction Series in Figure Story 5.5 and review the "Magmatic Differentiation" in this chapter.

19. The production of basalt can be achieved by the partial melting of
    A. gabbro.
    B. ultramafic rocks.
    C. a mixture of gabbro and oceanic sediments.
    D. rhyolite.
    Hint: Refer to Figures 5.11, 5.14, and 5.15.

20. Andesites—intermediate magmas—are typically associated with
    A. divergent plate margins, such as the mid-ocean ridges.
    B. fractures in the crust that allow magmas from the upper mantle to rise to the surface.
    C. Hot spots, such as Hawaii.
    D. convergent plate margins, such as the western edge of South America—the Andes.

21. An ophiolite suite contains which combination of rocks?
    A. granite, gneiss, sandstone, limestone, and shale
    B. deep-sea sediments, pillow basalts, gabbro, and peridotites
    C. dikes, sills, and plutons of peridotites
    D. plutonic rocks and surrounding sediments characteristic of a magma chamber

22. Which flowchart below characterizes the magmatic processes occurring at spreading centers?
    A. peridotite → decompression melting → basalt
    B. ophiolite → fluid-induced melting → basalt
    C. deep-sea sediments → fluid-induced melting → andesites
    D. basalt → decompression melting → peridotite

23. Which of the following hypotheses for the origin of granite is NOT reasonable?
    A. Mixing of magmas with different compositions
    B. Decompression melting within a magma plume
    C. Partial melting
    D. Melting of sedimentary and metamorphic rocks

24. Which set of conditions will result in basalt melting at the lowest temperature?
    A. Dry basalt at low pressure
    B. Dry basalt at high pressure
    C. Wet basalt at low pressure
    D. Wet basalt at high pressure

Igneous Rocks: Solids and Melts 65

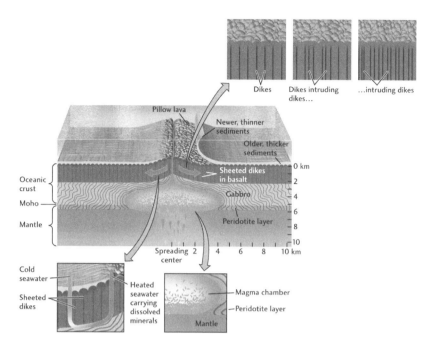

Figure 5.13. Magmatic processes at spreading centers involve decompression melting of peridotite to generate basalts.

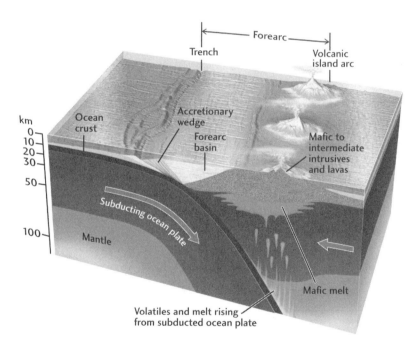

Figure 5.15. A schematic diagram of fluid-induced melting associated with subduction zones.

CHAPTER 6

Volcanism

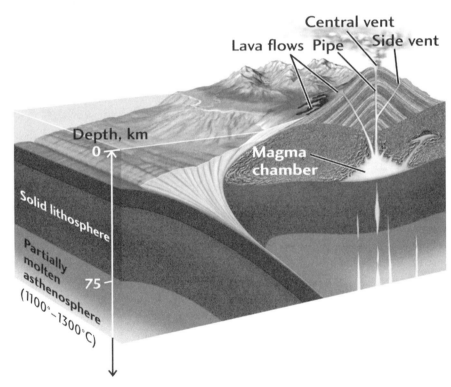

Figure 6.1. Simplified diagram of a volcanic geosystem.

Before Lecture

> ### Chapter Preview
>
> - **Why does volcanism occur?**
>   Brief answer: Volcanism occurs when magma rises buoyantly to the surface.
> - **What are the three major lava types and how do they relate to eruptive style and volcanic landforms?**
>   Brief answer: Mafic (basalt), intermediate (andesite), and felsic (rhyolite) lava types produce eruptions that range from relatively gentle basaltic flows to highly explosive rhyolitic eruptions. See Figure 6.9.
> - **How is volcanism related to plate tectonics?**
>   Brief answer: Volcanism is concentrated at convergent and divergent plate boundaries and hot spots. See Figure 6.19.
> - **What are some of the beneficial effects of volcanism?**
>   Brief answer: Volcanic processes create new ocean floor, our oceans and atmosphere, ore deposits, and geothermal energy.

Vital Information from Other Chapters

Review *How Do Igneous Rocks Differ from One Another* and *Igneous Activity and Plate Tectonics* in Chapter 5.

> ### Previewing Tip
>
> See **How to Study Geology** (in Part I of this Study Guide) for additional ideas on previewing.

During Lecture

One goal for lecture should be to leave class with a good set of answers to the preview questions.

- Focus on understanding how volcanoes work. Where the magma comes from (e.g., iron-rich ocean crust or silica-rich continental crust) has everything to do with the explosiveness of the eruption and the shape of the resulting volcanic landform.
- Focus particularly on understanding Figures 6.1 (how a volcano system works) and 6.9 (volcanic eruptive styles and landforms). It may be helpful to have these figures bookmarked and handy during lecture for quick referral. Do annotate the text figures with comments made by your instructor.

- Want ideas on taking a good set of notes? If you haven't already done so, read our discussion of note taking in **How to Be Successful in Geology** in Chapter 3 (Part I) of this Study Guide. You can use the **Note-Taking Checklist** before you go to lecture as a 1-minute reminder of what to do to improve your note-taking skills. After lecture use it as a quality check.

Note-Taking Tip: Abbreviations for Volcanism Terms

Use abbreviations to speed up your note taking:

Ig rock (igneous rock)
Strat V (stratovolcano)
Comp V (composite volcano)
HP (hot spot)
V (volcanism or volcano)
Sh V (shield volcano)
B (basalt)
Rhy (rhyolite or rhyolitic)
Mag (magma)

Use the following labels in the margin to draw attention to the material covered in that section:

TQ Use this abbreviation (test question) to indicate possible test questions.
?   Use a question mark beside areas where you felt lost or unclear about the material.

These flags will serve as a reminder that you need to follow up via further study or a conference with your instructor, tutor, or study partner.

After Lecture

Right after lecture, while the material is fresh in your mind, is the perfect time to review and improve your notes.

Check your notes: Have you...

☐ clearly identified important points? **Example:** You should have clear descriptions of the three lava types, information about the chemical composition and explosiveness of each type, and legible sketches of each landform discussed during the lecture.

☐ added sketches of the landforms associated with each eruptive style? Refer to Figure 6.9.

   Example: Compare the slope of a shield volcano with that of a composite volcano by drawing a simplified sketch of each showing how steep each volcano is and the relative area each is likely to cover. No fancy artwork is needed, just a steep inverted "V" for one and a flattened inverted "V" for the other. The shield volcano would be flattened and spread out and the composite (stratovolcano) steeper. Sketching provides a good check on how well you understand the differences.

☐ included a sketch of the basic features of a volcanic geosystem? Use Figure 6.1 as a reference.

☐ developed a table that summarizes the three lava types in terms of rock composition and associated landforms? Exercise 1 in **Practice Exercises and Review Questions** will help you complete the table and give you a perfect summary of Chapter 6.

Intensive Study Session

Ready to work? Set priorities for studying this chapter.

- **Practice Exercises and Review Questions.** Be sure to do Exercise 1. It gets to the key information you need to learn in this chapter.
- **Text.** Pay particular attention to Figures 6.1, 6.9, 6.10, 6.19, 6.22, and 6.23. These figure present the most important ideas in Chapter 6. Answer Exercise 4 and Thought Question 1 at the end of the chapter.
- **Web Site Study Resources**
  http://www.whfreeman.com/understandingearth

  Do the **Online Review Exercises:** *Where do volcanoes occur?* and *Finding the Volcanic Hot Spots.* Look at the volcanic landforms illustrated in the **Photo Gallery. Flashcards** will help you learn new terms. Complete the **Concept Self-Checker** and the **Web Review Questions** to assess your understanding of the chapter material. The **Geology in Practice** exercises explore the relationship between absolute plate motion and volcanic activity.

Motivation Tip

Chapter 6 contains material that should be of considerable interest to anyone living or planning to live in northwestern United States. There is information in this chapter that could, literally, save your life. Also illustrated are landforms in Hawaii and some of our western national parks (e.g., Mount Rainier, Crater Lake, and Yellowstone National Park). Quite possibly you have visited one of these places and wondered at their beauty.

Give yourself permission to enjoy this particularly interesting chapter. Begin your study session by just browsing through the artwork in Chapter 6 for material that interests you. Start reading wherever your interest takes you. Give yourself 15 or 20 minutes for just enjoying the interesting illustrations before you plunge into studying text and doing exercises.

Exam Prep

Materials in this section are most useful during your preparation for midterm and final examinations. The following **Chapter Summary** and **Practice Exercises and Review Questions** should simplify your chapter review. Read the **Chapter Summary** to begin your session. It provides a helpful overview that should refresh your memory.

Next, work on the **Practice Exercises and Review Questions.** In order to determine how you stand on mastery of this chapter, complete the exercises and questions just as you would a midterm. After you answer the questions score them. Finally, review any question that you missed. Identify and correct the misconception(s) that resulted in your answering the question incorrectly.

Chapter Summary

Why does volcanism occur?

- Volcanism occurs when molten rock inside the Earth rises buoyantly to the surface because it is less dense than the surrounding rock. Volcanism is a surface expression of magma generation within the Earth.

What are the three major lava types and how do they relate to eruptive style and volcanic landforms?

- Silicate lavas can be classified into three major types: mafic (basalt), intermediate (andesite), and felsic (rhyolite), respectively based on decreasing amounts of silica and increasing amounts of iron and magnesium.
- Eruption styles, volcanic deposits, landforms, and potential hazards are strongly linked to the chemical composition and gas content of the lava (refer to Exercise 1). Because basaltic lavas are relatively fluid and dry, they typically exhibit less explosive eruptions and erupt as lava flows. Rhyolite lavas are very viscous and usually wet. Therefore, they typically erupt very explosively as pyroclastic flows or form domes.

How is volcanism related to plate tectonics?

- There is a strong connection between major types of volcanism and crustal plate boundaries. Basaltic lavas occur at divergent plate boundaries and hot spots. The ocean crust is created by basaltic volcanism at the ocean ridge system. It is thought that basalt is generated by partial melting due to decompression of the ultramafic upper mantle. Basaltic, andesitic, and rhyolitic lavas erupt at convergent zones due to fluid-induced melting. The lavas generated along any particular convergent zone will depend in large part on what rocks are being subducted and melted within the overriding crust.

How does volcanism interact with human affairs?

- There are both benefits and hazards associated with volcanism. Geothermal heat is growing in importance for electric energy generation. Earth's oceans and atmosphere is thought to have condensed from volcanic degassing of our planet's interior. Volcanic dust and gases can impact global climate. Volcanic eruptions and associated mudflows can have disastrous impacts on a region and its people. Important ore-forming processes occur when hot water circulates through the magma chamber and surrounding rock.

Practice Exercises and Review Questions

Answers and explanations are provided at the end of the Study Guide.

Exercise 1: Lava types—Their Properties, Eruption Styles, Deposits, Landforms, Association with Plate Tectonics, and Hazards

This table will provide you with a very useful study guide for much of Chapter 6. Indeed the finished table makes an ideal summary of the chapter that should be very useful when you return to this unit in preparation for your midterm exam.

Fill in the blanks with the typical characteristics of each lava type. Keep in mind that different lavas exhibit a range of properties and behaviors. Give the best answer that generally characterizes each. Some answers have been provided as guidelines. Bullets mark information for you to complete.

Lava types	Basalt (mafic)	Andesite (intermediate)	Rhyolite (felsic)
Properties			
Eruption temperature	•	Intermediate	•
Silica content	•	Intermediate	High (≈70%)
Gas content	Low, up to a few percent	Variable	High (up to ≈15%)
Viscosity	Low-fluid magma	Intermediate	•
Typical flow velocity	0.7 to 30 m/minute	9 m/day	Less than 9 m/day
Typical flow length	10 to 160 km	8 km	Less than 1.5 km
Typical flow thickness	5 to 15 m	30 m	200 m
Eruption styles	Typically not very explosive.	•	•
Deposits	Flood basalt	Lava flow	Obsidian dome
	•	Dome	•
	•	Pyroclastic flow—	•
	•	tuff and welded tuff	
	•		
Landforms	•	Composite volcano	•
	•	Summit crater	•
	Cinder cone	Caldera	•
	Small caldera	Cinder cone	•
Association with plate tectonics	Hot spots	•	•
	•		
Hazards	•	Lava flow	Explosive blast
	•	Pyroclastic/ash flow	Hot gases
		Explosive blast	•
		Hot gases	•
		Mudflow	

Exercise 2: Volcanoes at plate tectonic boundaries

Complete this exercise by filling in the blanks adjacent to the list of volcanic areas with the correct match of type of volcano, characteristic magma type, and magmatic (plate tectonic) setting.

Use Figure Story 6.9, Figures 6.10, 6.18, and 6.25, and an atlas as a reference. Chapter 2 and web site links provided at http://www.whfreeman.com/understandingearth will also help you.

Sample answers are provided. Note that a volcanic area may have hybrid characteristics.

Volcano or volcanic area	Type of volcano (shield, composite, caldera)	Magma type (mafic, intermediate, felsic)	Magmatic (plate tectonic) setting—divergent, convergent, hot spot
Hawaii			
Tonga Islands	Composite	Intermediate & felsic	Convergent/subduction
Columbia Plateau		Mafic	Hot spot
Santorini (Thera), Greece	Caldera		
Mayon, Philippines			
Iceland			Divergent & hot spot
Yellowstone			
Krakatoa, Indonesia			
North Island, New Zealand			
Crater Lake, Oregon			
Japan			
Aleutian Islands, Alaska			
Mariana Islands			
Kilimanjaro, Africa			
Pinatubo, Philippines			
Katmai, Alaska	Composite & caldera		
Mount Rainier, Washington			
Tambora, Indonesia	Composite & caldera		Convergent
Vesuvius, Italy			

Review Questions

Answers and explanations are provided at the end of the Study Guide.

1. Lavas are almost always fine-grained rocks because
   A. the type of minerals they are made up of do not form large crystals.
   B. very little water is in the extruded magma.
   C. the lava crystallizes under pressure too low for large crystals to grow.
   D. they cool too rapidly for large crystals to grow.

2. Which extrusive rocks contain the most silica?
   A. Andesite
   B. Rhyolite
   C. Granite
   D. Basalt

3. During volcanic eruptions, the most common gas released is
   A. hydrogen.
   B. water.
   C. nitrogen.
   D. carbon dioxide.

4. A cloud of superheated steam and hot ash produced during a volcanic eruption, and moving rapidly parallel to the ground, is called
   A. a welded tuff.
   B. a pyroclastic flow.
   C. volcanic bombs.
   D. a cinder flow.

The longest word in the English language is supposedly **pneumonoultramicroscopicsilicovolcanoconios**—a lung disease caused by breathing in particles of volcanic matter or a similar fine dust.

5. A good example of shield volcanoes can be found in the
   A. volcanoes of northern California, Washington, and Oregon.
   B. Hawaiian Islands.
   C. Caribbean Islands, such as Mont Pelée on Martinique.
   D. volcanoes found along western South America.

6. Composite volcanoes (stratovolcanoes) are largely composed of
   A. basalt lava flows and basaltic cinders.
   B. pillow and pahoehoe lava flows.
   C. rhyolitic and intermediate lavas and pyroclastic flows.
   D. dikes and sills.

7. Compared to basalt, rhyolite lava flows are very thick and tend to form domes because rhyolite lava
   A. contains less gas than basalt lava and is more fluid.
   B. is richer in silica than basalt lava and is less fluid.
   C. cools more quickly than basalt lava.
   D. is less dense than basalt lava.

8. On a recent 3-day hike up a gently sloping mountain, your friends describe to you features they encountered. Frequently they crossed lava flows and fissures, and occasionally they had to detour around large cinder cones. Based on your friends' description, you tell them they were hiking on a
   A. caldera.
   B. composite volcano.
   C. gabbro pluton.
   D. shield volcano.

9. Composite volcanoes are commonly associated with which tectonic setting?
   A. Passive continental margins and deep-ocean basins
   B. Convergent plate boundaries, such as those of the circum-Pacific
   C. Ocean spreading centers, such as Iceland
   D. Transfer faults, such as the San Andreas fault

10. Of the following states, which is essentially all volcanic rock?
    A. Alaska   C. Oregon
    B. Hawaii   D. California

11. Typically, explosive volcanic eruptions are associated with
    A. basalt lavas.
    B. shield volcanoes.
    C. magmas that are poor in both silica and dissolved gases.
    D. magmas high in both silicand dissolved gases.

12. Shield volcanoes and composite volcanoes differ in shape based on
    A. the composition of the magma that forms each.
    B. the particular part of the ocean in which each forms.
    C. the latitude at which each forms.
    D. factors that are completely unknown to us at present.
    Hint: Refer to Figure 6.9.

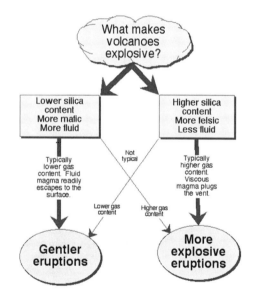

13. Calderas usually form
    A. due to molten material from the core coming very near the Earth's surface at a thin point in the crust.
    B. after a steam explosion, when magma comes into contact with abundant underground water.
    C. when an eruption literally blows off the top of a volcano.
    D. after large volumes of magma erupt, leaving a void in the magma chamber into which the superstructure of a volcano can collapse.
    Hint: Refer to Figure 6.10.

14. Your friends have described to you an eruption that took place at an undisclosed location. The lava they described merely flowed out of a fissure and spread rapidly over a large area. You would inform that person that the rock type being formed would most likely be
    A. granite.
    B. andesite.
    C. basalt.
    D. rhyolite.

15. You have been informed that an explosive volcanic eruption has taken place at an undisclosed location and that a huge nuée ardente (pyroclastic flow) flowed off the steep-sided volcano. You could respond that the magma type is likely to be
    A. basaltic.
    B. rhyolitic.
    C. ultramafic.
    D. gabbroic.

16. Where are andesitic volcanoes located?
    A. Where diverging plate boundaries occur
    B. At transform boundaries
    C. Along the mid-ocean ridge crest
    D. Along converging plate boundaries

17. A large resort located on a beautiful lake, and known for its hot springs, is experiencing a drop in business due to publicity about the recent swarm of small earthquakes in the area. The lake is actually located within a caldera, and beautiful rock towers and spires of weathered volcanic tuff are found all along the edge of the lake. As the director of the resort you're concerned about the change in business and the potential risk to your guests. What should you do?
    A. You cannot be worried because you know a volcano can't blow up on you.
    B. You know that earthquake swarms can be a precursor to volcanic eruptions, and that very explosive eruptions have happened at the place in the recent geologic past, so you decide that the resort should close until the situation is safe.
    C. Advertise the resort as the best place to see beautiful basalt lava fountains.
    D. Earthquakes have occurred occasionally along a nearby known fault, and there has been no historic volcanic eruptions, so you're not concerned.

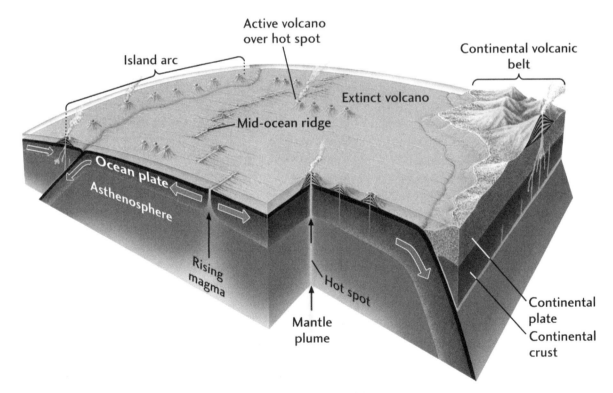

Figure 6.19. Volcanism is associated with plate tectonic boundaries and hot spots.

18. Which type of lava is most likely to erupt at the mid-ocean ridge? Refer to Figure 6.19.
    A. Basalt
    B. Andesite
    C. Rhyolite
    D. Diorite

19. Which type of lava is most likely to erupt at a hot spot? Refer to Figure 6.19.
    A. Basalt
    B. Andesite
    C. Rhyolite
    D. Diorite

20. Which type of lava is most likely to erupt at a convergent boundary along the edge of a continental plate? Refer to Figure 6.19.
    A. Basalt
    B. Andesite
    C. Rhyolite
    D. Diorite

21. Which of the following volcanic deposits can be formed from felsic lava?
    A. Pahoehoe
    B. Flood basalt
    C. Volcanic dome
    D. Shield volcano

22. If lava flows of progressively younger ages all erupted from a single large magma chamber, how would you expect their composition to have systematically changed?
    A. The lava flows will be progressively enriched in iron as they get younger.
    B. The lava flows will be progressively enriched in silica as they get younger.
    C. The lava flows will be progressively more mafic as they get younger.
    D. The lava flows will be progressively more fluid as they get younger.

23. Which of the following statements is true?
    A. Mafic rocks are richer in silica than felsic rocks.
    B. Mafic rocks crystallize at higher temperatures than felsic rocks.
    C. Mafic rocks are more viscous than felsic rocks.
    D. Mafic rocks tend to be lighter in color than felsic rocks.

24. Large volcanoes can potentially impact global climate when they erupt because they release
    A. geothermal heat.
    B. volcanic dust, sulfur, and carbon dioxide gas.
    C. lahars and lava flows.
    D. nitrogen and argon gases.

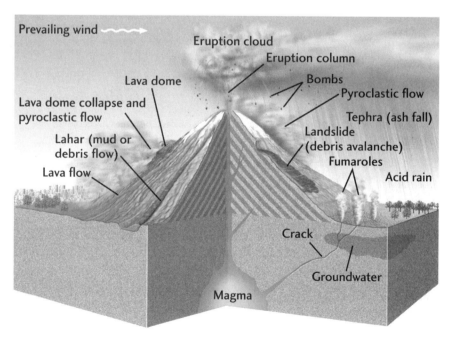

Figure 6.23. Some of the volcanic hazards that can kill people and destroy property.

CHAPTER 7

Weathering and Erosion

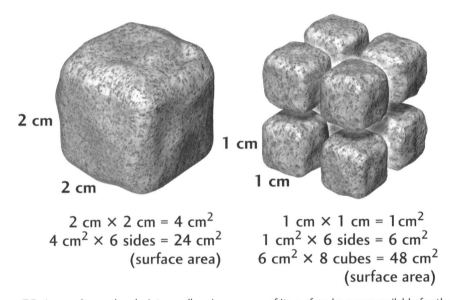

Figure 7.5. As a rock mass breaks into smaller pieces, more of its surface becomes available for the chemical reactions of weathering.

Before Lecture

Chapter Preview

- **What is weathering?**
  Brief answer: Weathering is the breakdown of rocks by chemical and physical processes.

- **How does chemical weathering work?**
  Brief answer: Water, oxygen, acids, and physical weathering facilitate chemical weathering reactions, which alter and break down minerals to form new minerals, oxides, salts, and release silica.
- **How does physical weathering work?**
  Brief answer: Physical weathering breaks rocks into fragments. Processes that aid physical weathering include chemical weathering, frost wedging, and plant roots.
- **How do soils form as products of weathering?**
  Brief answer: Soils form within rock materials due to chemical weathering. Soil formation is influenced by the composition of the rock, stability of the weathering surface (topography), time, and, most importantly, climate. Life processes and their byproducts are also important factors in soil formation and soil type.

*When we try to pick up anything by itself,
we find it entwined with everything else
in the universe.*

—JOHN MUIR

Vital Information from Other Chapters

- Section on *Rock-Forming Minerals* in Chapter 3 and Table 3.1.
- Section on *Sedimentary Rocks* in Chapter 4 and Figure 4.4.

Learning Warm-Up

Science and art are synergistic; that is, each enhances the other. For example, many aspects of nature we consider beautiful are the product of a weathering process. Before lecture, spend 5 minutes scanning the text photos for beautiful features and effects created by weathering. As you enjoy the photos you might ask yourself, "How did weathering processes create the colors and shapes in this photo?"

Man can live without gold, but not without salt.

—FALVIUS MAGNUS CASSIODORUS
A ROMAN POLITICIAN OF THE FIFTH CENTURY A.D.

During Lecture

As you take notes during this lecture be sure to get details on how weathering works.

- Focus on understanding specific chemical weathering processes, such as oxidation and dissolution. Be sure to distinguish clearly between the chemical processes (e.g., dissolution and oxidation) and the physical processes (e.g., frost wedging, exfoliation, etc.).
- Focus on understanding Figure 7.16, showing three soil profiles.

After Lecture

The perfect time to review your notes is right after lecture. If you wait even one day, most (80 percent) of what you heard will have disappeared from memory.

Check your notes: Have you...

- ☐ clearly identified the important points? **Example:** You should have clear descriptions of each type of chemical and physical weathering.
- ☐ added visual material? Chunking material is a good learning strategy. There are a lot of chemical and physical weathering processes in Chapter 7. It may be useful to make a list of processes that you can look at as a group all in one place in your notebook. Use a two-column format for this with processes in one column and details you need to remember in a second column.
- ☐ created a brief "big picture" overview of how weathering works? (See Exercise 1: Physical and Chemical Weathering.)
- ☐ added a clear comparison of the three soil types? (See Exercise 2: Soil Types.)

Intensive Study Session

Set priorities for studying this chapter. We recommend you give highest priority to activities that involve answering questions. Answering questions while using your text and lecture notes as reference material is far more efficient than reading chapters or glancing over notes. As always, you have three sources from which to choose questions.

- **Practice Exercises and Review Questions.** Begin with the **Practice Exercises and Review Questions**. Be sure to do the exercises as they will focus on the key information you need to learn in this chapter, namely how weathering works and the characteristics of the three soil types.
- **Text.** Answer *Exercises* 2, 4, 5, and 7 by using the **Media Link** category on the web site. Also complete *Thought Questions* 2, 3, and 8.
- **Web Site Study Resources**
  http://www.whfreeman.com/understandingearth

  Do the **Online Review Exercises:** *Understanding Soils* and *Mineral Stability*. The *Mineral Stability* exercise will help you review the susceptibility of common minerals to chemical weathering. Note that the susceptibility of common silicate minerals to chemical weathering is closely related to their silicate crystal structure. Quartz, a framework silicate, is very stable on the Earth's surface, whereas olivine with an isolated silica tetrahedral crystal structure is very susceptible to chemical weathering. Refer to Figure Story 3.11.

  The **Geology in Practice** exercises will allow you to apply what you learned about weathering to understanding the beautiful formations of Bryce Canyon National Park. Complete the **Concept Self-Checker** and **Web Review Questions**. Pay particular attention to the explanations for the answers.

Exam Prep

The time for taking your first geology midterm is close at hand. It is time to start thinking about how to organize your time. Here are a few tips that should make your exam prep more efficient.

Tips for Preparing for Exams

- Use the clues your instructor provides in lecture about what is important. Even when a department agrees on a common core of material (rare) each instructor carves out a course that is unique, has a particular character or flavor, and includes distinct areas of emphasis. Your instructor is the ultimate guide to the question "What is important?"
- Be sure you know the format of the exam. Multiple choice? True/false? Essay? Thought problems?
- Review your notes for material marked TQ (Test Question).
- Ask your instructor if exams are available from the previous semester. Review these to check the format of questions, what areas of content are stressed, and what types of problem solving are included. Don't make the mistake of assuming the same questions will be asked this semester.
- Be sure to attend review sessions if these are offered. Attending a review session will raise your exam score.
- If your class has tutors, preceptors, supplemental instruction leaders, or other peer helpers who have taken the class, ask for their suggestions about preparing for the exam.

Once you are clear about the nature of the exam, begin your review. Conduct the review in an orderly, systematic manner that ensures focused review of all the important material. Be sure to take a look at the Eight-Day Study Plan in Appendix A. The plan is a great model for orderly review.

Materials in this section are most useful during your preparation for midterm and final examinations.

The following **Chapter Summary** and **Practice Exercises and Review Questions** should simplify your chapter review. Read the **Chapter Summary** to begin your session. It provides a helpful overview that should refresh your memory.

Next, work on the **Practice Exercises and Review Questions.** In order to determine how you stand on mastery of this chapter, complete the exercises and questions just as you would a midterm. After you answer the questions score them. Finally, review any question that you missed. Identify and correct the misconception(s) that resulted in your answering the question incorrectly.

Chapter Summary

What is weathering?

- Physical weathering breaks rock into smaller pieces and chemical weathering alters and dissolves the minerals within the rock. The principal factors that influence weathering are the composition of the parent rock, topography (stability of the land surface), and climate. There is a positive feedback between physical and chemical weathering where one enhances the other if conditions are favorable. A good example of this positive feedback is soil formation. As physical and chemical weathering proceeds to alter a stable surface of rock material, a soil forms. The formation of soil promotes weathering by increasing the availability of moisture and producing acidic chemical conditions. Soils also promote the growth of plants, which aids both physical and chemical weathering.

How does chemical weathering work?

- How chemical weathering works is well illustrated by three examples. First, the chemical weathering of feldspars, which are the most abundant silicate mineral in the Earth's crust, illustrates how water with the help of carbonic acid can transform feldspars into clay minerals and dissolve silica and salts (cations). Refer to Figure Story 7.6. Second, the reaction of oxygen with the iron in ferromagnesium minerals, such as pyroxene, illustrates oxidation. Refer to Figure 7.8. Third, the reaction of calcite and other carbonate minerals making up limestone exemplifies the role naturally acidic water plays in dissolving rock.

How does physical weathering work?

- Physical weathering involves a variety of processes that break rock into fragments. Physical weathering is promoted by chemical weathering, which weakens grain boundaries within the rock. Physical weathering also promotes chemical weathering by increasing the surface area of the broken rock fragments. Frost wedging, mineral crystallization, and life processes play a major role in breaking apart rock.

How do soils form as products of weathering?

- Soils are a product of chemical weathering of rock that has remained in place for a period of time. Soil formation is most effected by climate. The composition of the parent rock, life processes, and topography are also important factors in soil formation. The three general types of soils are pedocals (arid climates), pedalfers (temperate climates), and laterites (tropical climates). Soils, water, and the air we breathe are the three most basic natural resources.

Practice Exercises and Review Questions

Answers and explanations are provided at the end of the Study Guide.

Exercise 1: Physical and chemical weathering

Fill in the blanks in the flowchart on page 81.

Exercise 2: Soil types

Fill in the blanks in the following table.

Climate	Soil type	Soil characteristics	Agricultural potential
Desert—warm and dry	*Pedocal*		
Temperate—moderate		*variable—clay rich*	
Tropical—warm and wet			*Lush vegetation is supported by an organic-rich A-horizon. Crops can be grown for only a few years before nutrients are depleted. Source of bauxite, an aluminum ore.*

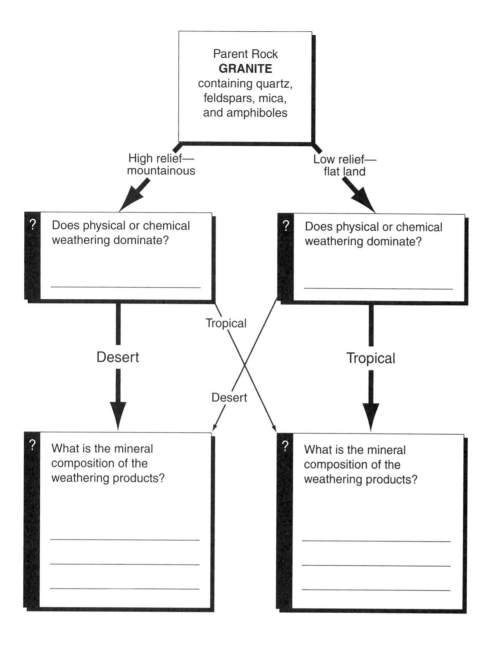

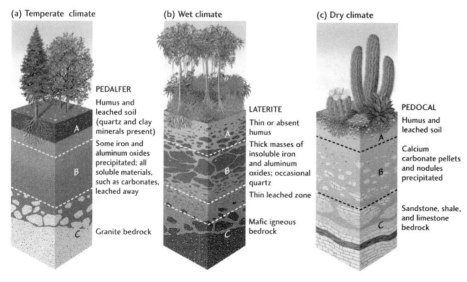

Figure 7.16. Soil profiles for temperate, dry, and wet climates.

Exercise 3: Soil formation in different regions

Given the parent rock and location, briefly characterize the major soil type (e.g., pedocal, pedalfer, or laterite) and the diagnostic characteristic of each.

A. Soil on a quartz sandstone in semiarid Kansas.
   Major soil type: _loose sand/sand dunes—minimum soil development._
   Characteristic(s): _The sandstone will break apart into quartz sand grains and_
   very little else.

B. Soil on a granite in semiarid Kansas.
   Major soil type: _____
   Characteristic(s): _____

C. Soil on a granite in warm and humid (temperate) Georgia.
   Major soil type: _____
   Characteristic(s): _____

D. Soil on a granite in semi-arid southern Arizona.
   Major soil type: _____
   Characteristic(s): _____

Review Questions

Answers and explanations are provided at the end of the Study Guide.

1. Of the following, which is NOT involved in the process of chemical weathering?
   A. Water
   B. Oxygen
   C. Carbon dioxide
   D. Nitrogen

2. The following products all result from chemical weathering EXCEPT
   A. feldspar.
   B. iron oxides.
   C. silica in solution.
   D. clay minerals.

3. Which of the following minerals does NOT chemically weather into a clay mineral?
   A. Muscovite
   B. K-feldspar
   C. Pyroxene
   D. Quartz

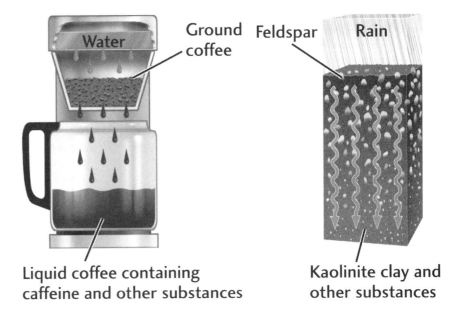

Figure 7.4. The process by which feldspar decays is analogous to the brewing of coffee.

4. Of the following materials, which one would make the longest lasting tombstone?
   A. Limestone
   B. Shale
   C. Granite
   D. Sandstone cemented with calcium carbonate

5. An example of chemical weathering is
   A. rusty streaks on a rock wall.
   B. angular blocks of rock rubble in the mountains.
   C. potholes in pavement.
   D. rocks wedged apart by tree roots.

6. Although water is an important agent of chemical weathering in its own right, it becomes more effective if small amounts of carbonic acid are present. Carbonic acid is formed when
   A. carbon from coal beds or graphite deposits is pulverized along a fault or fracture and then added to water.
   B. carbon dioxide from the atmosphere or organic decomposition is added to water.
   C. sulfur from coal-fired power plants is added to water.
   D. water comes in contact with the calcite in a limestone layer.

7. The potential for chemical weathering can be greatly enhanced by physical weathering because physical weathering
   A. increases the surface area of the rock particles.
   B. increases the availability of chemical agents.
   C. increases drainage and, thereby, reduces contact with water.
   D. increases the size of the rock particles.

8. Limestone and other carbonate rocks weather relatively fast in a _____ climate due to _____ weathering reaction.
   A. dry/oxidation
   B. dry/hydrolysis
   C. wet/physical
   D. wet/dissolution promoted by carbonic acid

9. Georgia soil, along with that from other warm, humid regions, is deep red in color. This color is due to
   A. iron oxides.
   B. quartz.
   C. feldspar.
   D. clay minerals.

10. Clay minerals, such as kaolinite, are a product of _____ weathering of _____ minerals and are a raw material for _____.
    A. chemical/silicate/pottery.
    B. physical/sulfate/cement.
    C. chemical/sulfide/asphalt.
    D. physical/carbonate/fertilizers.

11. Soluble mineral leaching in a soil occurs in which soil horizon?
    A. A-horizon
    B. B-horizon
    C. C-horizon
    D. All of the above

12. Which of the following climatic regions experiences the most rapid chemical weathering?
    A. Hot, low precipitation
    B. Extreme cold, low precipitation
    C. Hot, high precipitation
    D. Extreme cold, high precipitation

13. Which soil horizon tends to accumulate the most clay minerals and/or salts (such as calcium carbonate)?
    A. A-horizon
    B. B-horizon
    C. C-horizon
    D. R-horizon

14. Which of the following statements accurately describes a pedocal soil?
    A. They are typically enriched in aluminum in the B-horizon compared to laterite soils.
    B. They are typically enriched in calcium carbonate.
    C. They are typically enriched in iron and aluminum in the B-horizon compared to pedalfers.
    D. They are strongly leached soils and typically lack any significant amount of soluble material.

15. Generally, pedocal soils form in what climatic environments?
    A. Cool climates in heavily forested areas
    B. Warm climates with high precipitation
    C. Warm climates with low precipitation
    D. Tropical climates with extended periods of weathering

16. The soil production is often described as a "positive feedback process." Why?
    A. Because carbon dioxide in rainwater is used up by organisms, weathering of underlying rock is impeded.
    B. Because rainwater becomes more acidic as it percolates through the soil, weathering of underlying rock is promoted.
    C. Once a layer of soil is formed, the underlying rock is protected from further weathering.
    D. Plant growth reduces the potential for weathering and soil development.

17. Of the following minerals, the one most rapidly altered by chemical weathering would be
    A. mica (sheet silicate), such as biotite.
    B. amphibole (double-chain silicate), such as hornblende.
    C. pyroxene (single-chain silicate), such as augite.
    D. isolated silica tetrahedra mineral, such as olivine.
    Hint: Refer to Table 7.2.

18. The climate with the fastest rate of chemical weathering has
    A. low temperatures and low rainfall.
    B. high temperatures and high rainfall.
    C. high temperatures and low rainfall.
    D. low rates of mechanical weathering.

19. The potential for chemical weathering is _____ due to acid rain.
    A. neutralized   C. increased
    B. not effected   D. decreased

20. In the eastern United States, the widespread, good agricultural soil is called a
    A. pedalfer.   C. laterite.
    B. bauxite.   D. pedocal.

21. The generally great thickness of soils in tropical regions is due to
    A. spheroidal weathering of rock outcrop.
    B. rapid chemical weathering enhanced by high rainfall and high temperature.
    C. pressure release.
    D. action from the large variety of bacteria.

22. Which of the following processes generally results in a lateritic soil?
    A. Intense compaction of pedocal soils to form brick-hard material
    B. Intense chemical weathering
    C. Moderate to slight weathering
    D. Moderate weathering of very resistant lithologies

23. What happens to the quartz sand grains, chemically, in a calcite cemented sandstone that is undergoing moderate chemical weathering?
    A. They combine with water.
    B. They dissolve.
    C. They oxidize.
    D. Virtually nothing—they become grains of quartz sand.

24. Rock material that is _____ tends to result in the most fertile soils.
    A. not weathered at all   C. moderately weathered
    B. very weakly weathered   D. intensely weathered

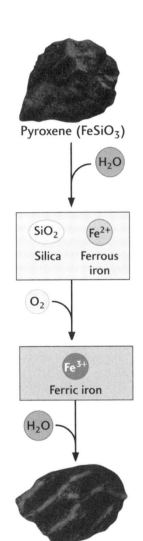

Figure 7.8. The general course of chemical reactions by which an iron-rich mineral, such as pyroxene, weathers in the presence of oxygen and water.

25. Physical and chemical weathering in the warm, wet climates of the Earth's surface will alter an exposed granite to
    A. quartz and feldspar sand.
    B. olivine sand.
    C. iron-rich soil.
    D. quartz sand and clay.

26. Even when weathering is intense, a soil may take _____ of years to form.
    A. tens
    B. hundreds
    C. thousands
    D. millions

    Hint: Refer to *Earth Policy* Box 7.1: *Soil Erosion*.

27. In the days of the Pharaohs of Egypt a cherished status symbol was the obelisk, a stone column decorated with hieroglyphs (designs carved into the stone—usually sandstone). In 1879 the obelisk of Thothmes III, from the temple of Heliopolis, Egypt, was moved to Central Park in New York City. Within about 60 years the hieroglyphs were barely visible on the obelisk, while its counterpart still standing in Egypt remains in nearly perfect condition in the desert sun for almost 4000 years. Why did the stone obelisk deteriorate so quickly when it was moved to New York City? **Hint:** Refer to Figure 7.1.

    _____
    _____
    _____
    _____

CHAPTER 8

Sediments and Sedimentary Rocks

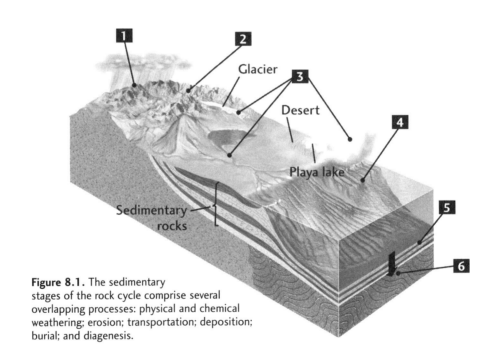

Figure 8.1. The sedimentary stages of the rock cycle comprise several overlapping processes: physical and chemical weathering; erosion; transportation; deposition; burial; and diagenesis.

Before Lecture

Chapter Preview

- **What are the major processes in the formation of sedimentary rocks?**
  Brief answer: weathering, erosion, transportation, deposition, burial, and diagenesis. Study Figure 8.1 and Figure Story 8.4.

- **What is the relationship between sediments and environments on the Earth's surface?**
  Hint: See Figure Story 8.4.
- **What are the major types of sediments and sedimentary rocks?**
  Brief answer: clastic (sandstone) vs. chemical (evaporites)/biochemical (limestone). Study Tables 8.3 and 8.4, Figure Story 8.4, and Figures 8.11 and 8.14.
- **How does plate tectonics relate to sedimentary rock?**
  Brief answer: Environments of deposition, composition and texture of sediments, and the shape and depth of sedimentary basins are influenced by plate tectonics. Review Figure 8.20.

Vital Information from Other Chapters

Review the rock cycle using Figure Story 4.9. Also review the processes of weathering and erosion in Chapter 7.

Note-Taking Tip

Some Features to Watch for in Sedimentary Rock Samples and Formations

In this lecture you will be introduced to a variety of sedimentary rocks, formations, and sedimentary environments. Because these will probably be new to you, it is smart to prepare. Spend a few minutes of your *preview time* becoming familiar with the most important sedimentary rock features:
- Bedding and other sedimentary structures: pages 171–174.
- Bedding sequences (stratigraphy): Figure 8.10 on page 174.
- Grain size: pages 176–179 and Figure 8.14: Clastic sedimentary rocks: conglomerate, sandstone, and shale.
- Mineral content and major groups of sandstones: Figure 8.15 on page 178.
- Chemical and biochemical sedimentary rocks: Figure 8.17.
- Fossils: Figure 8.18: Reefal limestone.

During Lecture

One goal for lecture should be to leave class with a good set of answers to the preview questions. To avoid getting lost in details keep the "big picture" in mind. Chapter 8 tells the story of how sedimentary rocks are created by the processes of the rock cycle. It may be helpful to have Figure 8.1 with you during class so you can refer to it during lecture.

Note-Taking Tip:

Use Abbreviations to Speed Up Your Note Taking

Bed → bedding	Lith → lithification
Carb → carbonate	Sed → sediment or sedimentary
Clast → clastic	Sed R → sedimentary rock
Chem → chemical	Carb sed → carbonate sediment
Biochem → biochemical	Sed base → sedimentary basin
Env → environment	Sed env → sedimentary environment

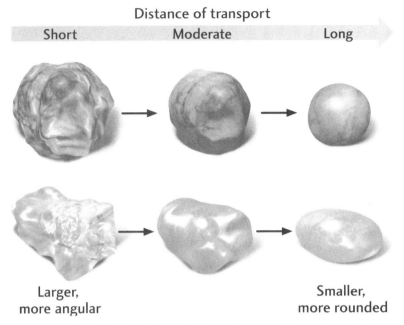

Figure 8.3. Transportation reduces the size and angularity of clastic particles. Grains become rounded and slightly smaller as they are transported, although the general shape of the grain may not change significantly.

After Lecture

Right after lecture, while the material is fresh in your mind, is the perfect time to review your notes. Review to be sure you got all the key points and wrote them down in a form that will be readable later.

Check your notes: Have you...

☐ included the chapter overview? Figure 8.1 is the "big picture." Consider adding your own version with the six sedimentary rock-forming steps included in your notes.

☐ developed a chart similar to the one in Exercise 1 to help you learn the sedimentary environments?

Intensive Study Session

Memory and Learning Tip: Chunking

It's easier to remember a list with three items on it than a list with 30 items on it. If you have 30 things to remember, the first thing you need to do is to think how you can divide the long list into a small number of shorter lists. Group similar items together, name the groups, and then associate the items on each list with the group to which it belongs. Instead of trying to remember 30 things you have reduced your task to remembering three.

Example: Look at Figure Story 8.4, which illustrates the different **sedimentary environments** on Earth. Note that while 11 sedimentary environments are discussed, these 11 are organized into just three kinds of environments: continental, shoreline, and marine. You will likely remember this information better if you chunk it into these three categories rather than trying to absorb the details of all 11 examples. The authors of *Understanding Earth* must agree. Notice that they have organized and color-coded the information to help you think in terms of just three general categories (continental, shoreline, marine) for sedimentary environments. Use these clues to study more effectively and improve your retention of the information.

In earlier chapters we have used the metaphor of building a house for studying geology. Chunking is a mental process that will help you construct a good frame. The idea is to group the important information. Later you hang details on this framework of basic information.

If you have sometimes wondered how some of your professors can rattle off tons of complicated material without even looking at their lecture notes, chunking (see box) provides an explanation. Over the years they have chunked (connected) more and more bits of information about geology together. Chapter 8 provides a great opportunity for you to develop your memory efficiency. Use the clues and aids identified below to group the details of sedimentary rocks into a format that is easy to remember.

Study the Text: Focus on Chunking the Information into a Solid Overview

- First, get an overview of the processes involved in forming sedimentary rocks. Study Figure 8.1 and commit the six steps in the figure to memory.
- Next, work on Figure Story 8.4 on page 169, which illustrates common *sedimentary environments* on Earth. Note that while there are 11 sedimentary environments discussed, these 11 are organized into just three kinds of environments. For now focus most of your attention on understanding those three and don't get lost in the details of all 11.
- Move on to *diagenetic processes* on pages 174–175. Figure 8.11 provides a great visual overview. It chunks lots of information together. Here's a little memory trick: You can recall all the information in this figure story by remembering just two words: "process" and "sediments." Here's how it works. First, notice that Steps 1–5 describe the three simple processes (pressure, compaction, and cementation) that are a part of diagenesis. Next, look at Step 6. Observe how pressure transforms different sediments into different kinds of rocks with very different textures. There are four sediments (mud, sand, gravel, and organic matter). You have two short lists to remember: the list of three processes and the list of four sediments. Quite likely if you can remember the sediments (mud, sand, gravel, organic matter) you can easily guess the rock that each turns into. You can commit the entire table to memory by memorizing a very short two-word list: *processes (3)* and *sediments (4)*. The rest of the table should fall out by association. Try it!
- Move on to *classification* on page 176. There are many details to sedimentary rock classification. Notice, however, that you can reduce all these details to two basic categories:

    (1) clastic sedimentary rocks (Table 8.3 and Figure 8.14) and (2) chemical/biochemical sedimentary rocks (Table 8.4 and Figure 8.17). **Hint:** If you are a visual or kinesthetic learner, it may help to work up your own simplified version of these tables.

If you want to test your knowledge of the chapter, the materials below can be used for self-testing and applying your new knowledge to some very interesting geologic problems.
- **Practice Exercises and Review Questions.** At the very least do Exercise 1. It gets to the key information you need to learn in this chapter.

- **Text.** Read *Earth Policy 8.1* on pages 182–183. Study this to learn more about the formation of coral reefs. This is a must if you like to scuba dive! Answer Exercises 1, 2, and 5. Also, work on Thought Questions 3, 5, 7, and 12. When you visit the Grand Canyon in Arizona, you will be able to understand the geology better and impress your traveling partners after completing these exercises and questions.
- **Web Site Study Resources**
  http://www.whfreeman.com/understandingearth
  Go to the web site where you can try your hand at identifying and reconstructing the interesting story told by sedimentary rocks using the **Geology in Practice** exercises.

  The following **Online Review Exercises** will also provide you with an opportunity to practice using information and skills highlighted in this chapter: *Identify the Sedimentary Environment, Identify the Common Sedimentary Environment,* and *Identify the Depositional Environment.* Use the Flashcards to help you learn the new terminology and use the **Concept Self-Checker** and **Web Review Questions** to assess your understanding of the chapter.

Exam Prep

Materials in this section are most useful during your preparation for midterm and final examinations.

The following **Chapter Summary** and **Practice Exercises and Review Questions** should simplify your chapter review. Read the **Chapter Summary** to begin your session. It provides a helpful overview that should refresh your memory.

Next, work on the **Practice Exercises and Review Questions.** In order to determine how you stand on mastery of this chapter, complete the exercises and questions just as you would a midterm. After you answer the questions score them. Finally, review any question that you missed. Identify and correct the misconception(s) that resulted in your answering the question incorrectly.

Chapter Summary

What are the major processes in the formation of sedimentary rocks?

- Weathering and erosion produce the clastic particles and dissolved ions that compose sediment. Water, wind, and ice transport the sediment to where it is deposited. Burial and diagenesis harden sediments into sedimentary rocks. Refer to Figure 8.1.

What is the relationship between sediments and environments on the Earth's surface?

- Sedimentary environments are often grouped into three general locations on the continents, near the shoreline, or in the ocean.
- Factors that influence the sediments in these environments include: (1) kind and amount of water, (2) topography, and (3) biological activity.
- Sedimentary structures and fossils provide information about the agent of transport (water, wind, or ice) and environment of deposition for the sediment.

What are the major types of sediments and sedimentary rocks?

- The two major types of sediments are clastic and chemical/biochemical. Clastic sediments are formed from rock particles and mineral fragments. Chemical and biochemical sediments originate from the ions dissolved in water. Chemical and biochemical reactions precipitate these dissolved ions from solution.
- Classification and the name of clastic sediments and sedimentary rocks (conglomerate, sandstone, and shale) are based primarily on the size of the grains within the rock. The classification and name of chemical and biochemical sediments (evaporites, limestones, and dolostone) in sedimentary rock are based primarily on their composition.

How is sediment transformed into hard rock?

- Burial and diagenesis transforms loose sediments into hard rock. Burial promotes this transformation because buried sediments are exposed to increasingly higher pressures and temperatures. Diagenesis involves many physical and chemical processes. For example, cementation is an important chemical diagenetic change in which minerals are precipitated in the open spaces within sediments, forming cements that bind together clastic sediments.
- Figure 8.11 summarizes how diagenetic processes produce changes in composition and texture.

How does plate tectonics relate to sedimentary rock?

- Sedimentary environments, the composition and texture of sediments, and the geometry of the basins in which sediments accumulate are all related to plate tectonic settings. For example, the formation of a new ocean basin by rifting along a divergent plate boundary creates a rift basin and ultimately a stable continental margin and ocean basin (refer to Figure 8.20).

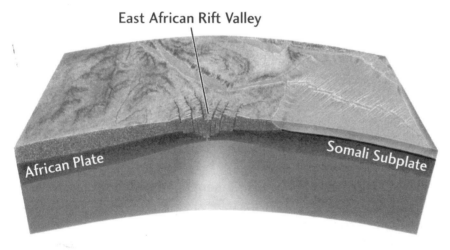

Figure 8.21. Rifting and plate separation within a continent illustrates the development of three major sedimentary basins: continental rift valley, young linear ocean basin, and mature ocean basin.

Practice Exercises and Review Questions

Answers and explanations are provided at the end of the Study Guide.

Exercise 1: Common sedimentary environments

Using Figure 8.1 and Figure Story 8.4 as a guide, fill in the blanks in the following table with the clastic or chemical sediment, (e.g., sand, silt, mud, salts, carbonate, and peat) that best matches the following environment of deposition.

Environment of deposition	Sediment
Alpine or glacial river channel	
Dunes in a desert	
Flood plain along a broad meander bend	
River delta along a marine shoreline	
Continental shelf	
Deep sea adjacent to a continental shelf	
Shoreline beach dunes	
Tidal flats	
Organic reef	

Exercise 2: Grain sizes for clastic sedimentary rocks

Using the terms for *sediments* and *rocks* listed in Table 8.3, fill in the blanks with the appropriate name of the sediment and rock that matches the typical particle size of the *Common Object* given in the following table. **Hint:** A few answers have been filled in as a reference.

Grain size	Common object	Sediment	Rock type
Coarse ↑	Football to bus	*Boulder gravel*	
	Plum or lime		*Conglomerate*
	Pea or bean		
	Coarse-ground pepper or salt		*Sandstone*
	Fine-ground pepper or salt		
↓ Fine	Talcum powder or baby powder		

Exercise 3: Clastic and chemical sediments and sedimentary rocks

Given the descriptive statements on the left side, fill in the blanks with the appropriate sediment type and rock type. A list of common sediment types are provided below for your reference. Use the following terms for *sediment type* and *sedimentary rock* to fill in the following table.

Sediment types	Sedimentary rocks			
biochemical	arkose	dolostone	limestone	sandstone
clastic	chert	evaporite	peat	shale
chemical	conglomerate	graywacke	phosphorite	siltstone

Statement	Sediment type	Sedimentary rock example
Composed largely of rock fragments		
Precipitated in the environment of deposition		
Important source of coal		
Often formed by diagenesis	Chemical	Dolostone and phosphorite
Formed from abundant skeleton fragments of marine or lake organisms, such as coral, seashells, foraminifers		
Produced by physical weathering		
Produced from rapidly eroding granitic and gneissic terrains in an arid or semiarid climate		

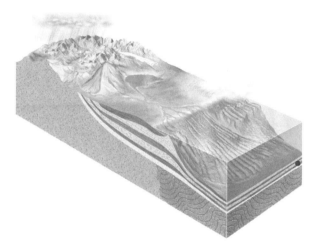

Figure 8.11. (Part 1) Sediments are buried, compacted, and lithified at shallow depths in the Earth's crust.

Figure 8.15. Mineralogy of four major groups of sandstone.

Review Questions

Answers and explanations are provided at the end of the Study Guide.

1. Which of the following rock groups includes only clastic sedimentary rocks?
    A. Dolomite, gypsum, and limestone
    B. Cherts, sandstone, and shale
    C. Dolomite, coal, and limestone
    D. Shale, sandstone, and conglomerate

2. Which sequence of rock names has the correct arrangement in order of decreasing particle diameters?
    A. Conglomerate, shale, sandstone
    B. Shale, siltstone, sandstone
    C. Sedimentary breccia, shale, sandstone, claystone
    D. Conglomerate, sandstone, claystone

3. The grains in a sandstone may include
    A. rock fragments.
    B. quartz.
    C. feldspar.
    D. all of the above.

4. Sedimentary rocks are produced through the following sequence of events:
    A. erosion, weathering, transportation, deposition, burial, and diagenesis.
    B. weathering, erosion, transportation, burial, diagenesis, and deposition.
    C. erosion, weathering, deposition, transportation, and cementation.
    D. weathering, erosion, transportation, deposition, burial, and diagenesis.

> **Test-Taking Tip:**
>
> When taking a multiple-choice test, treat each alternative answer as a true-false question. Rule out any alternative that is false. For example, in Question 4, A and C are false because erosion must follow weathering. B is false because deposition must proceed burial and diagenesis. Therefore, D must be the correct answer.

5. Dolomite is the primary mineral found in dolostone. It is formed by
    A. foraminifera extracting minerals from seawater.
    B. diagenetic alteration of calcite.
    C. direct precipitation in lake water.
    D. coral reef exposure to direct sunlight.

6. What are the two most important diagenetic processes that transform loose sediments into hard sedimentary rocks?
    A. Compaction, cementation
    B. Transportation, burial
    C. Erosion, transportation
    D. Uplift, erosion

7. What are two chemical or biochemical sedimentary rocks formed by diagenetic process?
    A. Dolostone, phosphorite
    B. Limestone, chert
    C. Sandstone, limestone
    D. Siltstone, graywacke

8. Where are most sediments deposited?
    A. On the continental shelf and adjacent ocean floor
    B. In lakes
    C. Along streams
    D. In deserts

9. In which of the following would you least expect to find cross-bedding?
    A. Sand dunes
    B. Delta sediment
    C. River bar deposits
    D. Evaporites

10. The most widespread environment of deposition for carbonates in the world today is
    A. the deep sea; e.g., the Arctic Ocean.
    B. the tidal flat environment; e.g., the Mississippi Delta.
    C. in river channels.
    D. a warm, shallow-water marine environment; e.g., Florida Keys.

11. A sandstone made of pure quartz grains is likely to be deposited
    A. along the shore of a continent.
    B. in the deep sea.
    C. in river channels.
    D. along the edges of a coral reef and atolls.
    Hint: Refer to Figure 8.15.

12. Chemical weathering is most dominant in
    A. warm, dry climates.
    B. warm, wet climates.
    C. cool, dry climates.
    D. cool, wet climates.
    E. at the shoreline.

> **Study Tip**
>
> Review Chapter 7: Weathering and Erosion.

13. The coast of Alaska is known for its high mountainous relief, active volcanoes, and glaciers that reach to the sea. What types of sediments would you expect to be commonly deposited off shore from this landscape?
    A. Arkose
    B. Carbonate sand
    C. Quartz sand
    D. Graywacke
    Hint: Refer to Figure 8.15 and the text above this figure.

14. When a granite is subject to intense chemical weathering, it most likely will result in a sediment composed of
    A. feldspar and clay.
    B. quartz and calcium carbonate.
    C. quartz and clay.
    D. quartz and feldspar.

15. Given that feldspar is the most abundant silicate mineral in the crust of the Earth, the most common sedimentary rock is
    A. sandstone.
    B. conglomerate.
    C. limestone.
    D. shale.

16. A coarse sandstone with asymmetrical ripples and small scale cross-bedding is exposed in a cliff between layers of siltstone above and gravels below. What is the environment of deposition for the coarse sandstone layer?
    A. Beach
    B. River channel
    C. Lake
    D. Off shore

17. Reefs and atolls are built by coral and algae
    A. on subsiding oceanic volcanoes and continental margins.
    B. on islands in the middle of lakes.
    C. in the deep ocean floor and later uplifted to sea level.
    D. where dolostone is transformed to limestone.

18. As seawater evaporates, precipitation of soluble salts occurs in the following order:
    A. first halite, then carbonates, and finally calcium sulfate.
    B. first calcium sulfate, then halite, and finally carbonates.
    C. first iron oxide, then quartz, and finally peat.
    D. first carbonates, then calcium sulfate, and finally halite.

19. Mechanisms by which plate tectonic processes produce sedimentary basins are
    A. rifting, thermal sag, and flexure of the lithosphere.
    B. by heating and compressing the crust.
    C. by weathering and erosion.
    D. diagenesis and turbation.

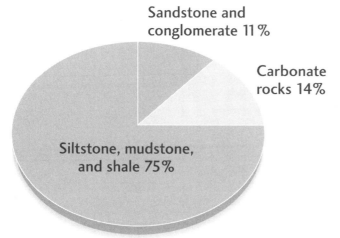

Figure 8.13. The relative abundance of the major sedimentary rock types.

CHAPTER 9

Metamorphic Rocks

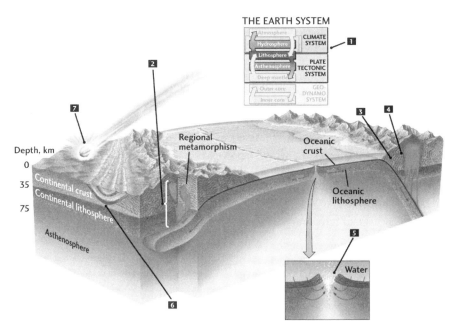

Figure 9.3. The main types of metamorphism and where they occur.

Before Lecture

Chapter Preview

- **What causes metamorphism?**
  Brief answer: Metamorphism—alteration of preexisting rocks in the solid state—is caused by increases in pressure and temperature and by reaction with chemical components introduced by migrating fluids. See Figure 9.1.

- **What are the various kinds of metamorphism?**
  Brief answer: Regional and contact metamorphism are the most common. See Figure 9.3 and Figure Story 9.4.
- **What are the chief types of metamorphic rocks?**
  Brief answer: Metamorphic rocks fall into two major textural classes called foliated with minerals oriented in some preferred direction, such as the grain in wood, and nonfoliated with no preferred mineral orientation. See Table 9.1.
- **How is metamorphism linked to plate tectonics?**
  Hint: Refer to Figure 9.3.

Vital Information from Other Chapters

Review *Metamorphic Rocks* and *The Rock Cycle* in Chapter 4. These sections are short and well worth your review.

During Lecture

Keeping up with a fast-speaking lecturer can be quite a challenge.
- Take as many notes as you can.
- Don't stop writing when you get confused or if you want to ponder a concept. You can do that later.
- If you miss something leave a space so you can fill it in later.

Note-Taking Tip:
Metamorphic Rock Abbreviations to Speed Up Your Note Taking

Met rock → metamorphic rock
Sed rock → sedimentary rock
Sl → slate or slaty
Sl cleave → slaty cleavage
Shl → shale
Shst → schist
Fol → foliated
Unfol → unfoliated
Gran → granular
Reg M → regional metamorphism
Con M → contact metamorphism

Feel free to make up your own abbreviations. The important thing is to develop shorthand that is meaningful to you, while being quick, easy to use, and easy to remember.

After Lecture

Review your notes right after lecture while material is fresh in mind. Here is some metamorphic rock material to add to your notes.

Check your notes: Have you...

- ☐ added key figures to focus your attention? For this chapter the most important pieces are Figure 9.1 (pressure and temperature), Figure 9.3 (types of metamorphism), Figure Story 9.4 (foliation), and Table 9.1 (classification).
- ☐ added helpful sketches? Suggestion for Chapter 9: *Metamorphic rocks are classified in part by texture. Draw simple sketches of textures (slaty cleavage, phyllite, schist, gneiss) to help you remember the grades of metamorphism. Close study of the photo at the beginning of the chapter and Figure Story 9.4 will help you see how to do this.*
- ☐ added a summary of the graphs? Chapter 9 contains some important information in graphic form. Review P-T Figures 9.2, 9.8, 9.9, and 9.10. Note that all show how different *grades* of metamorphism result from increasing pressure and temperature. Can you summarize all these figures on a single page of your notes? What does your summary sketch tell you about the formation of metamorphic rocks?
- ☐ created a brief "big picture" overview of this lecture (using a sketch or written outline)? Suggestion for Chapter 9: *Figure 9.3 is key to understanding the tectonic settings that drive metamorphism. Sketch a simplified version that clearly shows six geologic settings for metamorphism. Write a caption for this figure in your own words.*

Intensive Study Session

We recommend you give the highest priority to activities that involve answering questions. Answering questions while using your text and lecture notes as reference material is far more efficient than reading chapters or glancing over notes. As always you have three sources from which to choose questions.

- **Practice Exercises and Review Questions.** Use the **Practice Exercises and Review Questions** at the end of this chapter. Exercise 1 will help you sort out the kinds of rock discussed in this chapter. Now that you have studied the three rock types in detail, it's time to integrate the information. Exercise 2 will help you rise above the details and gain an overview by comparing igneous, sedimentary, and metamorphic rocks.
- **Text.** For this chapter, the most important pieces are Figure 9.1 (pressure and temperature), Figure 9.3 (types of metamorphism), Table 9.1 (classification), and Figure 9.10 (plate tectonics). Complete Exercises 1, 6, 8, and 9 and Thought Questions 5, 7, and 8 at the end of the chapter.
- **Web Site Online Review Exercises and Study Tools.**
  http://www.whfreeman.com/understandingearth
  Complete the **Concept Self-Checker** and **Web Review Questions**. Pay particular attention to the explanations for the hints and answers. **Flashcards** will help you learn new terms. The **Geology in Practice** exercise provides an opportunity to practice identifying metamorphic rocks and reviewing geologic settings in which they occur. Complete the **Online Review Exercises**: *What Happens Where?, Metamorphism & Plate Tectonics, Create a Metamorphic Rock,* and *Before & After: Metamorphic Rocks* to review and reinforce what you have learned in the text.

Exam Prep

Materials in this section are most useful during your preparation for midterm and final examinations.

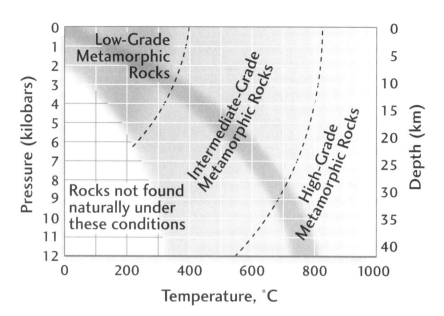

Figure 9.2. Temperatures, pressures, and depths at which low- and high-grade metamorphic rocks form. The dark band shows common rates at which temperature and pressure increase with depth over much of the continents.

The following **Chapter Summary** and **Practice Exercises and Review Questions** should simplify your chapter review. Read the **Chapter Summary** to begin your session. It provides a helpful overview that should refresh your memory.

Next, work on the **Practice Exercises and Review Questions.** In order to determine how you stand on mastery of this chapter, complete the exercises and questions just as you would a midterm. After you answer the questions score them. Finally, review any question that you missed. Identify and correct the misconception(s) that resulted in your answering the question incorrectly.

Chapter Summary

What causes metamorphism?

- Metamorphism is the alteration in the solid state of preexisting rocks, including older metamorphic rocks. Increases in temperature and pressure and reactions with chemical-bearing fluids cause metamorphism. Metamorphism typically involves a rearrangement (recrystallization) of the chemical components within the parent rock. Rearrangement of components within minerals is facilitated by (1) higher temperatures that increase ion mobility within the solid state; (2) higher confining pressure that compacts the rock; (3) directed pressure associated with tectonic activity that can cause the rock to shear (smear), which orients mineral grains and generates a foliation; and (4) chemical reactions with migrating fluids may remove or add materials and induce the growth of new minerals.

What are the various kinds of metamorphism?

- An overview of six types of metamorphism is provided in Figure 9.3. Regional metamorphism (associated with convergent plate boundaries), contact metamorphism (caused by the heat from an intruding body of magma), and seafloor metamorphism (caused by seawater percolating at mid-ocean spreading centers) are the three most common types of metamorphism within the Earth's crust. Other types of metamorphism are low-grade, high-grade, and shock metamorphism.

What are the chief types of metamorphic rocks?

- Metamorphic rocks fall into two major textural classes: the foliated (displaying a preferred orientation of minerals, analogous to the grain within wood) and the nonfoliated. The composition of the parent rock and the grade of metamorphism are the most important factors controlling the mineralogy of the metamorphic rock. Metamorphism usually causes little to no change in the bulk composition of the rock. The kinds of minerals and their orientation do change. Mineral assemblages within metamorphic rocks are used by geoscientists as a guide to the original composition of the parent rock and the conditions during metamorphism. Refer to Figure Story 9.7.

- Mineral assemblages in metamorphic rocks provide a basis for reconstructing the conditions that caused metamorphism and understanding more about the associated geologic setting. Figure 9.8 and Table 9.2 summarize major minerals of metamorphic facies. Figure Story 9.7 and Figures 9.9 and 9.10 illustrate how geologists study and interpret metamorphic rocks and reconstruct the conditions that formed them.

How is metamorphism linked to plate tectonics?

- Regional and high-pressure metamorphism occurs at convergent plate boundaries. Refer to Figures 9.4 and 9.10.
- Seafloor metamorphism occurs at spreading centers.
- Contact metamorphism occurs in a variety of plate tectonic settings where magma bodies are generated.

The important thing is not to stop questioning.

—ALBERT EINSTEIN

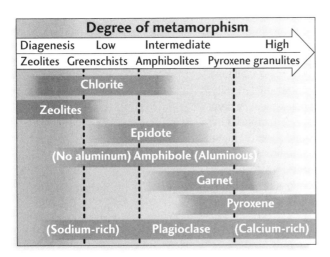
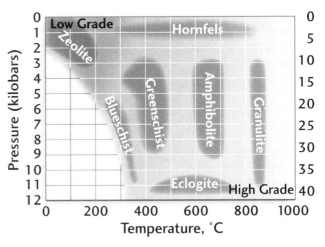

Figure 9.8. (a) Changes in the mineral composition of mafic rocks, metamorphosed under conditions ranging from low grade to high grade. (b) The metamorphic facies of mafic rock types.

Practice Exercises and Review Questions

Answers and explanations are provided at the end of the Study Guide.

Exercise 1: Classification of metamorphic rocks based on texture
Complete the table by filling in the blanks.

Parent rock	Metamorphic rock	Texture (foliated/granoblastic)
Shale		*Foliated*
Quartz-rich sandstone		
	Granulite	
Granite		
Limestone		
	Hornfels	
	Amphibolites and greenstones	
	Migmatite	

Study Tip: Putting It All Together

Now you have been introduced to all three major rock types: igneous, sedimentary, and metamorphic. This is a good time to assemble what you have learned into a comparison chart. Doing so is an excellent way to ensure you remember details about these rock types. Exercise 2 will help you do this.

Exercise 2: Comparing igneous, sedimentary, and metamorphic rocks.
Complete the table by filling in the blanks.
  Note that there may be more than one reasonable answer for some blanks.

Major mineral composition	Texture	Rock type (igneous, sedimentary, metamorphic)	Rock name (granite, sandstone, marble)
Calcium carbonate	*Nonfoliated*		
Quartz, K & Na feldspar, mica, and amphibole	*Phaneritic*		
Clay	*Fine-grained clastic*		
Pyroxene, calcium feldspar, and olivine			*Basalt*
Quartz	*Nonfoliated*		
Pebbles and cobbles of a variety of rock types			
Fragments of seashells and fine mud		*Sedimentary*	
Quartz, muscovite, chlorite, and garnet	*Metamorphic*	*Schist*	

Review Questions

Answers and explanations are provided at the end of the Study Guide.

1. When existing rocks undergo metamorphism, they become changed by
   A. the weathering process at or near the surface.
   B. color and hardness.
   C. melting and crystallization from that melt.
   D. the application of heat and pressure.

2. Metamorphic rocks exposed at the surface are mainly products of processes acting on rocks
   A. near the Earth's surface.
   B. at depths ranging from the upper to lower crust.
   C. within the mantle.
   D. within the center of continents.

3. Generally there are two types of metamorphic rocks:
   A. regional and contact.
   B. foliated and granoblastic.
   C. compacted and cemented.
   D. clastic and porphyritic.

4. Foliated metamorphic rocks typically occur in association with regional metamorphism because
   A. the orientation of rocks involved in regional metamorphism favors the development of foliation.
   B. the rock is softened by heat and squeezed by compressive forces.
   C. the parent rock is the correct type to produce foliation.
   D. the pressure is very low, which allows foliation to develop.
   Hint: Refer to Figure Story 9.4.

5. Foliation develops
   A. perpendicular to compressive forces.
   B. parallel to compressive forces.
   C. due to high water content.
   D. due to low temperatures and pressures.

6. Some metamorphic rocks are distinguishable from igneous and sedimentary rocks because their constituent grains often
   A. interlock, forming a continuous mosaic.
   B. have quite different chemical compositions.
   C. tend to be lined up in a preferred direction.
   D. are rounded and cemented together.

7. Chemical compositions of metamorphic rocks are determined by the
   A. pressures to which they have been subjected.
   B. effects of both temperature and pressure.
   C. temperature to which they have been raised.
   D. composition of the original rocks and fluids that percolate through the rock during metamorphism.

8. Metamorphism affects
   A. only older igneous rock.
   B. any younger igneous and metamorphic rock.
   C. only older sedimentary rock.
   D. any older igneous, sedimentary, or metamorphic rocks.

9. Which metamorphic sequence correctly shows increasing grain size?
   A. schist → gneiss → phyllite → slate
   B. slate → phyllite → schist → gneiss
   C. gneiss → phyllite → slate → schist
   D. phyllite → slate → gneiss → schist

10. Starting with the lowest temperature zone, which series of index minerals is characteristic of increasing metamorphic grade?
    A. Chlorite, biotite, garnet, sillimanite
    B. Garnet, chlorite, biotite, kyanite
    C. Biotite, garnet, chlorite, sillimanite
    D. Chlorite, biotite, sillimanite, kyanite, garnet
    Hint: Refer to Figure Story 9.7.

11. On the Moon, the major cause of metamorphism is
    A. burial.
    B. subduction.
    C. meteor impacts.
    D. very cold temperatures.

12. Of the metamorphic rocks listed below, which one is formed at the highest temperature?
    A. Slate
    B. Schist
    C. Phyllite
    D. Gneiss
    Hint: Refer to Figure Story 9.7.

13. The difference between a gneiss and a granite is that the gneiss
    A. has a different bulk chemical composition.
    B. shows a distinct foliation.
    C. has a different mineral composition.
    D. is generally less coarse-grained.

14. Schist and slate are distinguishable in that
    A. schist is fine-grained, whereas slate is coarse-grained.
    B. schist is foliated, whereas slate is not foliated.
    C. slate is fine-grained, whereas schist is coarse-grained.
    D. slate is foliated, whereas schist is not foliated.

15. Regional metamorphism is found in association with
    A. lava flows.
    B. hot springs.
    C. very low pressures.
    D. subduction zones and cores of mountain ranges.

16. If gneiss or another metamorphic rock is heated to a degree that it begins to melt,
    A. the quartz, K-feldspar, and Na-rich plagioclase would start to melt first.
    B. Na-Ca plagioclase, biotite, and minerals (such as garnet) would melt first, leaving a residue rich in felsic minerals.
    C. all the minerals in the rock would start to melt at essentially the same temperature to form a magma of the same composition of the gneiss.
    D. the ferromagnesian minerals would start to melt first.
    Hint: Refer to Figure Story 5.5.

17. As magma intrudes into a host or country rock (the preexisting rock that comes into contact with the intrusion) is transformed into a new rock. What do we call this process?
    A. Regional metamorphism
    B. Contact metamorphism
    C. Recrystallization
    D. Schistosity formation

18. In the *Geology in Practice* Exercise 1: *Gravels to Metaconglomerate,* oval quartz pebbles apparently become converted into cigar-shaped features within a metaconglomerate. How would you explain the transformation of the oval pebbles into cigar-shaped features?
    A. The cigar-shaped features formed as larger quartz cobbles were tumbled in a stream channel before metamorphism.
    B. As the oval quartz pebbles became exposed to elevated temperatures, they softened and stretched out in response to directed pressure. Some of the quartz may have also recrystallized along directions perpendicular to the directed pressure.
    C. Quartz is an index mineral for high-grade metamorphism associated with burial of sedimentary rocks. Therefore, this conglomerate must have been uplifted from near the bottom of the crust.
    D. The conglomerate melted and the large cigar-shaped features are large quartz crystals.
    Hint: The bottom of Figure Story 5.5 will provide an important clue.

19. You are on a summer backpacking trip in Alaska with friends and find an outcrop of mica schist with large garnet porphyroblasts (see the sample shown in Figure 9.6). Your friends quickly scramble to collect some garnet crystals. They then ask you about the conditions under which the beautiful garnet crystals formed. Your response to their query is:
    A. Garnet-bearing schists are formed from iron-rich magmas that solidify underground.
    B. Garnet is an index mineral for low-temperature and pressure metamorphism associated with meteor impact craters. So, we must be within an ancient impact crater.
    C. Garnet commonly occurs in mica schists and is an index mineral for intermediate to high-grade metamorphism, associated with regional metamorphism. This rock may have at one time been in the roots of a huge mountain.
    D. Garnets only occur in eclogites. Therefore, this rock must have oriented at the base of the crust and extensive uplift and erosion has occurred in this region to exposed it.

20. While studying some metamorphic rock samples for an undergraduate research project, you discover evidence that the rocks were exposed to high-grade metamorphic conditions. Further study reveals that the garnets within the rock show a history of both prograde followed by retrograde P-T paths. You check the field notes from the geologist who collected the samples and are not surprised to find mention of ophiolites in the region (refer to Figure 5.12). You conclude from this information that metamorphic rocks formed when

   A. continents collided.
   B. magma intruded into a volcano associated with a subduction zone.
   C. a meteorite hit the Earth.
   D. hydrothermal fluids altered rock in an area with numerous hot springs and geysers.

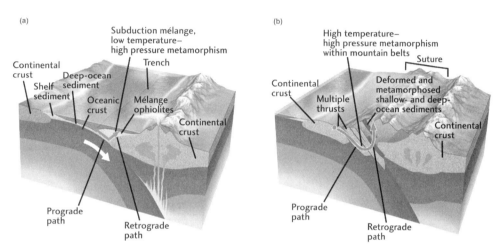

Figure 9.10. P-T paths and rock assemblages associated with (a) ocean-continent plate convergence and (b) continent-continent plate convergence.

CHAPTER 10

The Rock Record and the Geologic Time Scale

*Strata, no matter how different they may appear,
are of the same age if they contain the same assemblage of fossils.*

—William "Strata" Smith
(1769–1839)

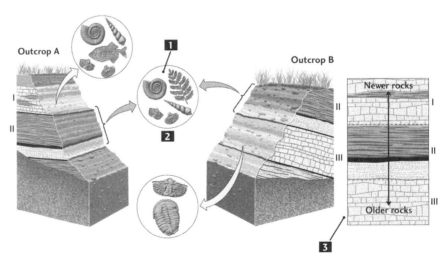

Figure 10.5. William Smith could piece together the sequence of rock layers of different ages containing different fossils by correlating outcrops found in southern England. In this example, formations I and II were exposed at outcrop A, and formations II and III at outcrop B.

Before Lecture

Time-Management Tip:
Something is always better than nothing

How much time should you devote to preview? Obviously, the more time spent previewing, the better. However, even a brief (5- or 10-minute) preview session just before lecture will produce noticeable results. Arrive 10 minutes early for lecture. Use the time to preview the chapter for that day's lecture. Even if you only have a minute or two you can read the preview questions and brief answers and gain a rough idea of what the chapter is about. The time just before lecture is precious, because whatever you preview will remain in short-term memory and help you better understand lecture.

Chapter Preview

- **How can the relative ages of rocks be determined from rock outcrops?**
  Brief answer: The principles of superposition and cross-cutting relationships provide a basis for establishing the relative age of a sequence of rocks at an outcrop. For example, in Figure 10.3 and *Earth Issues* 10.1, the youngest sedimentary rock layers are on top and the oldest layers are on the bottom. In Figure 10.9, magma intrusions and a fault are younger than the rock through which they cut.
- **How can the relative ages for rock outcrops at two or more locations be determined?**
  Brief answer: Stratigraphic and fossil succession and radiometric dates of rock units provide a basis for establishing how rock outcrops at different localities may be related to each other (correlate), even if they are 100s or 1000s of miles apart. See Figure 10.5.
- **What is the Geologic Time Scale and how is it calibrated?**
  Brief answer: The Geologic Time Scale is the internationally accepted reference for Earth's geologic history. Using relative and absolute dating methods, geologists have calibrated (divided) Earth history into four eons: Hadean, Archean, Proterozoic, and Phanerozoic. Because more evidence is available for the most recent eon, the Phanerozoic, it has been possible to divide it more finely into eras, periods, and epochs. See Figures 10.12 and 10.17.
- **How are the Geologic Time Scale and geochronological methods, such as radiometric dating, applied to geologic problems?**
  Brief answer: We understand Earth's history to the degree to which we can place the record of geologic events in time. The Geologic Time Scale is the accepted standard for how geologic time is subdivided.

Vital Information from Other Chapters

- Review Figure 1.12 (The ribbon of geologic time from the formation of the solar system to present).
- *Where We See Rocks* section in Chapter 4.
- *Atomic Structure of Atoms* in Chapter 3 before reading about radiometric dating methods.

*Time is simply nature's way of keeping everything
from happening at once.*

—GRAFFITI ON A WALL

During Lecture

Here's an overview that should help you get a good set of notes for this lecture.

- **Big Picture.** The "big picture" for this lecture is the entire Geologic Time Scale. Geologic time is wondrously huge. It is so expansive that, at first, it seems incomprehensible. Your lecturer will describe the 4.5-billion-year expanse of geologic time, and may use examples and exercises designed to help you grasp geologic time.

- **New Terms.** It is difficult to talk about the Geologic Time Scale without referring to its time intervals. So you may feel a bit barraged with new terms: epochs, periods, eras, eons, Holocene, Pleistocene, Pliocene, etc. To avoid getting lost keep a copy of Figure 10.12 close at hand. Refer to Figure 10.12 to check terms as needed. Better yet, look over this chart before lecture. Note how different sized chunks of time are used for different eons: huge chunks (eons) for events early in Earth's history when there is less geological evidence for what happened then and smaller chunks (epochs) for intervals of time closer to the present. Because more evidence is available for the most recent eon, the Phanerozoic, it has been possible to divide it more finely: into eras, periods, and even epochs. See Figures 10.12 and 10.17.

- **The Succession of Geologic Events.** You will work through how the *relative* ages of rocks are determined. By the end of lecture you will be able to use two basic principles (superposition and cross-cutting) to determine the relative age of a sequence of rocks such as that shown in Figure Story 10.11.

- **Dating Methods.** The last chunk of information in this lecture deals with *absolute* dating of rocks. Carbon-14 has a short half-life. It works for dating younger samples of tissue attached to bone, charcoal, and wood because these materials all contain carbon. Other isotopes (uranium-238, potassium-40, rubidium-87) have much longer half-lives and are used to date rocks that are geologically older. Table 10.1 shows how half-life is related to the effective dating range of each method.

After Lecture

The perfect time to review your notes is right after lecture. The following checklist contains both general review tips and specific suggestions for this chapter.

Check your notes: Have you...

- ☐ shown clearly how superposition and cross-cutting features can be used to sequence rock units and geologic events? Hint: A simple sketch is the best way to show this.
- ☐ shown clearly the characteristics of (1) an unconformity and (2) an angular unconformity? Check Figures 10.6, 10.7, and 10.8 against your notes. Again add sketches if you need to.

Intensive Study Session: Strategies for learning the Geologic Time Scale

The two big tasks for this study session are 1) to master the skill of dating outcrops using the principles of superposition and cross-cutting and 2) to learn the Geologic Time Scale.

Begin your work on outcrops with *Exercise 1: Determining the Succession of Geologic Events*. Many of the other exercises and review questions will allow you to test your skills further. Refer to your notes and relevant text figures to help you.

The second task is to learn the Geologic Time Scale. Does that sound like a challenge? Unless you happen to have spent your summer working on an archeology dig (tossing around terms like Paleocene, Eocene, Oligocene, etc....) the epochs of the Cenozoic may look like a steep memory curve. Here are four different strategies you can use to help you learn the Geologic Time Scale.

1. **Marker Events** are simply interesting things that happened: animals or plants that evolved, creatures that dominated the earth, large extinction events. Look at Figure 10.12. Select some marker events you already know about. **Example:** Can you guess one of the periods during which dinosaurs were dominant? The movie *Jurassic Park* has made this an easy question to answer. When did complex life begin? Find some other marker events of particular interest to you. You may find yourself surprised at how early some events occurred. Marker events will help you remember the Geologic Time Scale. Try **Practice Exercise 3:** *Marker Events for the Geologic Time Scale*.

2. **Logical Chunks.** Group the information into short lists you can remember. Study the groupings of the time scale with Figure 10.12 in front of you. Learn it as a series of short lists. Understand the following logic:
   - **Eons** are the biggest time chunks. There are only four (Hadean, Archean, Proterozoic, Phanerozoic) to remember. Only the most recent (Phanerozoic Eon) is broken down further. Hadean sounds like Hades (which is Hell in ancient Greek mythology), not a bad description of the young planet with its molten surface and asteroids crashing into it.
   - Eras are the next biggest time chunks. You only have to learn eras for the Phanerozoic. There are only three Phanerozoic Eras to remember: Old Life, Middle Life, and New Life. Think of it this way first, you can tack on the Greek stems later (see #3 below).
   - Periods are next. All three eras are further divided into periods.
   - Epochs are the smallest divisions of geologic time. You only have to learn epochs for the most recent era (Cenozoic or New Life). All epochs of the Cenozoic end in "cene" (for Cenozoic).

3. **Word Stems.** Word stems are clues to meaning. Greek and Latin stems are used a great deal by scientists. You can look them up in any good dictionary. A few helpful stems for the Geologic Time Scale include:

   Eras:
   Paleo- = Greek: "old"
   Meso- = Greek: "middle"
   Ceno- = Greek: "new"
   -zoic = Greek: "life"

   Epochs: Don't worry about the middle epochs for now. Just remember the first and last ones.
   Paleo = Greek: "old"
   Pleisto = Greek: "much." Remember there was *much* ice in the Pleistocene.
   Holo = Greek: "recent." Remember: The Holocene is the present or most recent epoch.

   Note that all epochs of the Cenozoic epoch end in "cene."

4. **Mnemonic (catch phrase).** When there are long lists of unfamiliar terms to learn (such as the epochs of the Cenozoic) many learners find it helpful to make catch phrases to help them remember. Try this. Do **Practice Exercise 4:** *Geologic Time Scale Mnemonic*.

Web Site Online Review Exercises and Study Tools

http://www.whfreeman.com/understandingearth

Take the **Concept Self-Checker** quiz and pay particular attention to the feedback for answers. *Flash Cards* will help you learn new terms. Work with *Field Relationships for Relative Time Dating* and the *Geologic Time Scale Review* Online Review Exercise.

*No vestige of a beginning,
no prospect of an end.*

—JAMES HUTTON
(1726–1797)

Exam Prep

Materials in this section are most useful during your preparation for midterm and final examinations. The following **Chapter Summary** and **Practice Exercises and Review Questions** should simplify your chapter review. Read the **Chapter Summary** to begin your session. It provides a helpful overview that should refresh your memory.

Next, work on the **Practice Exercises and Review Questions.** In order to determine how you stand on mastery of this chapter, complete the exercises and questions just as you would a midterm. After you answer the questions score them. Finally, review any question that you missed. Identify and correct the misconception(s) that resulted in your answering the question incorrectly.

Chapter Summary

How can the relative ages of rocks be determined from rock outcrops?

- The principles of superposition and cross-cutting relationships provide a basis for establishing the relative age of a sequence of rocks at an outcrop. Using these two principles, geologists can order (what happened first, second, third, so on) the geologic events represented by the rocks and geologic features in rock outcrops. The principle of original horizontality for sedimentary layers provides a basis for identifying sequences of sedimentary rocks affected by tectonic forces after they were deposited.

How can the relative ages for rock outcrops at two or more locations be determined?

- To reconstruct the geologic history of the Earth, geologists need to correlate the geologic events represented by rocks at one locality with the geologic events represented by rocks at other localities. The stratigraphic and fossil succession and radiometric dates of rock units provide a basis for establishing how rock outcrops at different localities may be related to each other, even if they are 100s or 1000s of miles apart.

What is the Geologic Time Scale and how is it calibrated?

- The Geologic Time Scale is the internationally accepted reference for the sequence of events represented by Earth's rock record. It was constructed over about the last 200 years by geologists using mainly fossils, superposition, and cross-cutting relationships to establish the relative ages for thousands of rock outcrops around the world. In about the last sixty years, the Geologic Time Scale has been calibrated using radiometric methods.

How is the Geologic Time Scale and geochronological methods, such as radiometric dating, applied to geologic problems?

- We understand Earth's history to the degree to which we can place the record of geologic events in time. The Geologic Time Scale is the accepted standard for how geologic time is subdivided. *Earth Issues* 10.1 provides an example.

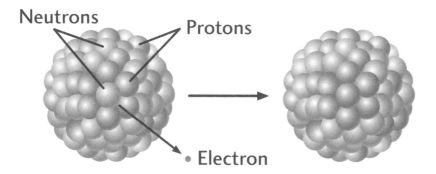

Figure 10.13. The radioactive decay of rubidium to strontium.

Practice Exercises and Review Questions

Answers and explanations are provided at the end of the Study Guide.

Exercise 1: Determining the succession of geologic events

The block diagram below illustrates the geology of an area in Argentina. Answer the following questions regarding the geological history of this area. Circle the correct answer when a choice is provided.

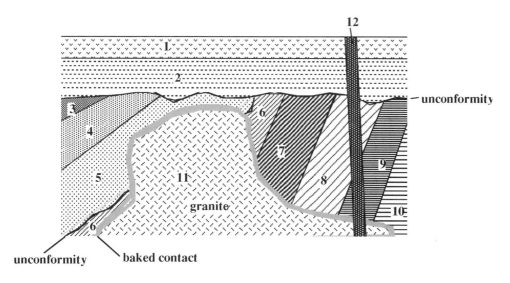

Notes:
Unit 2 contains clasts of units 3 through
Unit 5 contains clasts of units 6 through
Units 5 through 10 are baked along their contacts with unit
Units 1, 2, 8, 9, and 11 are baked along their contacts with dike

A. Which unit is the youngest rock in this area?   1 2 3 4 5 6 7 8 9 10 11 12

B. Which unit is the oldest rock in this area?   1 2 3 4 5 6 7 8 9 10 11 12

C. 1. Is unit 3 older than unit 12?   Yes   No   Not possible to know

   2. Explain the logic you used to answer the above question (C, Question 1).
_____
_____

D. 1. Is unit 1 younger than unit 11?   Yes   No   Not possible to know

   2. Explain the logic you used to answer the above question (D, Question 1).
_____
_____

E. 1. In an attempt to further work out the geologic age relationships in this area, samples of units 1, 11, and 12 were collected for radiometric dating to determine the ages of these igneous rocks. The resulting counts of radioactive parent atoms and daughter atoms are listed in the table below.

Rock Unit	# Parent atoms	# Daughter atoms
1	500	500
11	250	750
12	750	250

Using this information, unit 1 is

1. older than unit 12 but younger than unit 11
2. younger than unit 12 but older than unit 11
3. older than unit 11 but younger than unit 12

  2. Explain the logic you used to answer the above question (E, Question 1).
_____
_____

Exercise 2: Ordering geologic events

In the illustration below, a geologic outcrop reveals three layers of sedimentary rocks, one fault, and a single igneous dike intrusion.

A. Using this illustration, order the sequence of geologic events from youngest to oldest.

Youngest _____
_____
_____

Oldest _____

B. Briefly describe the geologic history represented by the rock sequence illustrated above.
_____
_____
_____
_____
_____

Exercise 3: Marker events for the geologic time scale

A. Enter each event, listed below, on the line in the appropriate eon box in which the event happened. When possible order your listing so that the oldest is at the bottom of the list and the youngest is on top.

B. Fill-in the names of eras, periods, and epochs in the correct sequence from oldest at the bottom to youngest on top. Refer to Figures 1.12 and 10.12 to complete this exercise.

<u>Significant (Marker) Events in Earth History</u>

Dinosaur extinction event	Major phase of continent formation completed
Earliest evidence of life	Moon forms
End of heavy bombardment of the Earth	First nucleus-bearing cells develop
Evolutionary Big Bang	Oxygen buildup in atmosphere
Humans evolve	

Eon	Era	Period	Epoch
Phanerozoic *Humans evolve* _____		*Quaternary* *Tertiary*	*Holocene* *Pleistocene* _____ _____ _____ _____
	Mesozoic	*Jurassic* _____	
	_____	*Pennsylvanian* _____ _____ *Ordovician* _____	
Proterozoic _____ *First nucleus-bearing cells develop*			
Archeon _____ _____			
Hadean _____ _____ *Earth accretion begins*			

Exercise 4: Geologic Time Scale mnemonic

Construct a mnemonic device for remembering the Geologic Time Scale names. The first letter of each word must match the first letter of the corresponding period or epoch in the proper order. You may use your native language, but be careful not to mix up the words when you do so.

Examples (refer to Figure 10.12 for the Geologic Time Scale):

Here's a good mnemonic for the Periods of the Geologic Time Scale:

Chronically Overworked Student Decks Monotonous Physics Professor To Justify Contradictory Test Questions.

Here's a good mnemonic for the Epochs of the Cenozoic:

Please Eat Our Mushroom Pot Pie Hot.

Now it is your turn to invent your own memorial mnemonic to help you remember the Geologic Time Scale.

Review Questions

Answers and explanations are provided at the end of the Study Guide.

1. The principle of superposition holds that for any unfolded series of sedimentary layers,
   A. overlying strata extend over a broader area than any layers beneath them.
   B. the layer at the top of the pile is always younger than those beneath it.
   C. sediments generally accumulate in the vertical sandstone-shale-limestone sequence.
   D. the layer at the top of the pile is always older than those beneath it.

2. From youngest to oldest, the correct sequence of eras dividing the Phanerozoic eon is:
   A. Paleozoic, Cenozoic, Mesozoic.
   B. Mesozoic, Cenozoic, Paleozoic.
   C. Mesozoic, Paleozoic, Cenozoic.
   D. Cenozoic, Mesozoic, Paleozoic.

3. The epochs of the Tertiary period progress from oldest to youngest in which sequence?
   A. Eocene, Oligocene, Paleocene, Miocene, Pliocene
   B. Paleocene, Eocene, Oligocene, Miocene, Pliocene
   C. Paleocene, Eocene, Miocene, Pliocene, Oligocene
   D. Paleocene, Eocene, Oligocene, Miocene, Pliocene, Pleistocene

4. From oldest to youngest, the correct order of periods for the Paleozoic era is:
   A. Cambrian, Ordovician, Devonian, Silurian, Mississippian, Permian, Pennsylvanian.
   B. Cambrian, Ordovician, Silurian, Devonian, Mississippian, Pennsylvanian, Permian.
   C. Cambrian, Silurian, Ordovician, Pennsylvanian, Mississippian, Devonian, Permian.
   D. None of the above.

5. Sandstone and shale rock layers immediately below and above an angular unconformity imply a history of
   A. erosion, deposition, deformation, and erosion.
   B. erosion, deformation, deposition, and erosion.
   C. deformation, erosion, deposition, and deformation.
   D. deposition, deformation, erosion, and deposition.
   Hint: Refer to Figure 10.8.

6. From the diagram, one can infer age limits for rock layer 3 of
   A. between 34 and 60 million years.   C. less than 20 million years.
   B. between 30 and 60 million years.   D. more than 60 million years.

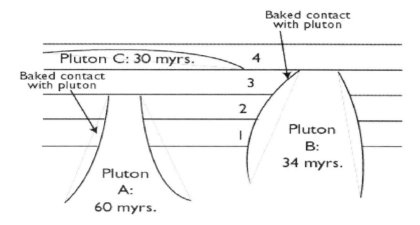

7. What makes the isotopes of a given element different from each other?
   A. Their atomic numbers
   B. Their number of electrons
   C. Their number of neutrons
   D. Their number of protons

8. Naturally occurring _____ that decay(s) into other materials at known rates can be used to estimate the actual age of a rock.
   A. organic matter   C. radioactive elements
   B. minerals   D. silicon

9. The Phanerozoic time is divided into three eras: (1) the interval of old life, (2) the interval of middle life, and (3) the interval of modern life. These intervals correspond to the (from oldest to youngest)
   A. Archean, Mesozoic, and Paleozoic.
   B. Paleozoic, Precambrian, and Proterozoic.
   C. Precambrian, Cambrian, and Neocambrian.
   D. Paleozoic, Mesozoic, and Cenozoic.

10. Which radiometric dating method would be most effective to determine the age of charcoal at an archeological site?
    A. Rubidium-strontium   C. Uranium-lead
    B. Radiocarbon   D. Potassium-argon
    Hint: Refer to Table 10.1 in your textbook.

11. Only geologically young materials can be dated using radioactive C-14 isotopes because
    A. the decay rate varies widely over time.
    B. they have a very short half-life.
    C. within decades all the C-14 is decayed away.
    D. they are a very rare isotope.

12. If the half-life of some radioactive element is one billion years, and a mass of rock originally contained 1000 atoms of the radioactive element, how many atoms of the radioactive element would be left after three billion years had passed?
    A. 500 atoms
    B. 250 atoms
    C. 125 atoms
    D. No radioactive atoms

13. Small pieces of charcoal from an ancient ruin yield a carbon-14 date of 3000 years. This age best represents the approximate interval of time that has elapsed since
    A. a fire burned the wood.
    B. humans inhabited the ruin.
    C. humans cut the wood.
    D. the wood died.

14. Radiometric dates have been attached to the Geologic Time Scale by the determination of
    A. radiometric ages of igneous rocks younger and older than sedimentary formations.
    B. radiometric ages of shales.
    C. radiometric ages of fossil skeletons.
    D. radiometric ages of metamorphosed sediments.

15. For the most part, radiometric dates for rocks only represent the last time the rock
    A. crystallized from a magma or was metamorphosed.
    B. was eroded.
    C. became cemented.
    D. was deposited.

16. One method that geologists use to study buried sediments and unconformities is
    A. seismic stratigraphy.
    B. radiometric stratigraphy.
    C. depositional stratigraphy.
    D. metamorphic stratigraphy.
    Hint: Refer to Figure 10.10.

17. If a rock is heated by metamorphism and the daughter atoms generated by the decay of the radioactive parent atoms migrate out of a mineral that is subsequently radiometrically dated, the date will be _____ the actual age.
    A. younger than
    B. older than
    C. the same as
    D. None of the above.

18. Which sample of basalt in the diagram is likely to yield the most accurate K/Ar radiometric date?

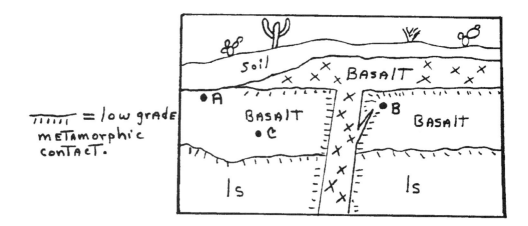

    A. A
    B. B
    C. C

19. A layer of conglomerate contains cobbles of an igneous rock. One of the cobbles was dated radiometrically at 35 million years old. From this radiometric date, the conglomerate layer can be inferred to be
    A. more than 35 million years old.
    B. less than 35 million years old.
    C. 35 million years old.
    D. none of the above.

20. In the illustration below, what is the most recent geological event depicted?

    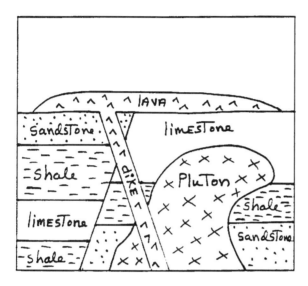

    A. Eruption of the lava
    B. Faulting
    C. Intrusion of the pluton
    D. Deposition of shales, sandstones, and limestones

21. Pieces of charcoal were found in a paleosoil layer covering an ancient fire pit with stone tools including arrowheads and axes. The charcoal was radiometrically dated using carbon-14 and yielded an age of approximately 10,500 years before present. What can you infer about the age of the archeological site?
    A. The archeological site is about 10,500 years old.
    B. The archeological site is younger than 10,500 years.
    C. The archeological site is older than 10,500 years.
    D. The age of the archeological site is unresolved but may be older than 10,500 years.

CHAPTER 11

Folds, Faults, and Other Records of Rock Deformation

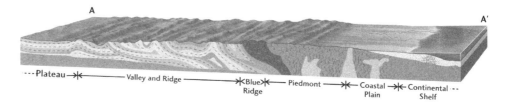

Figure 11.21. The Valley and Ridge province of the Appalachian Mountains is the eroded remnant of a folded mountain belt.

Before Lecture

Before you attend the lecture be sure to spend some time previewing Chapter 11. For an efficient preview, use the questions below. **Chapter Preview** questions constitute the basic framework for understanding this chapter. Preview works best if you do it just before lecture. With the main points in mind you will understand the lecture better. This in turn will result in a better and more complete set of notes.

> **Study Tip**
>
> Rock Deformation is a particularly visual lecture. Slide material on folds and faults can be confusing if you have never seen these geologic features before; therefore, be sure to preview the figures before attending this lecture. For an overview start with Figure Story 11.6. Notice how three kinds of force produce three kinds of faults if the material is brittle. What happens if the material is ductile or plastic?

How much time should you devote to preview? Obviously, more time is better than less. But even a brief (5- or 10-minute) preview session just before lecture will produce a result you will notice. For a refresher on why previewing is so important see **How to Be Successful in Geology** in Chapter 3 (Part I) of this Study Guide.

Chapter Preview

- **How do rocks deform (bend and break)?**
  Brief answer: Rocks typically break (fault) when temperature is low, burial is shallow, and the force is applied quickly. Rocks typically bend (fold) when temperature is higher, burial is deeper, and the force is applied over a long period. See Figure Story 11.6.
- **What geologic features are produced when rocks deform?**
  Brief answer: Folds, faults, and joints are common geologic structures produced when rocks are deformed. See Figure Story 11.6 and Figure 11.16.
- **What do geologic structures produced by rock deformation tell geologists about the geologic history of a region?**
  Brief answer: The type(s) of folds, faults, joints, and their spatial orientation provides geologists with clues for deciphering the kinds of forces affecting a region over time. See Figure 11.22.

Vital Information from Other Chapters

After each lecture you need to thoroughly master the concepts covered. You need to do this before you attend the subsequent lecture. The ideas of geology are like a stack of boxes. Each new idea rests on all ideas (boxes) stacked beneath it.

Another look at Chapter 4 would serve you well. Because there is a strong connection between deformation and metamorphism the following sections in Chapter 9 are recommended for review:

Figure Story 9.4 and the section of text entitled *Plate Tectonics and Metamorphism*. Finally, be sure you understand the Principle of Superposition in Chapter 10.

Web Site Preview

http://www.whfreeman.com/understandingearth
The **Online Review Exercise:** *Tectonic Forces in Rock Deformation* introduces you to the basic terminology and concepts covered in this chapter.

During Lecture

One goal for lecture is to leave class with a good set of answers to the preview questions.

- To avoid getting lost in details, keep the "big picture" in mind. Chapter 11 tells the story of three kinds of forces (compressive, tensional, and shearing) and how geologists find evidence of these forces in rock structures (folds and faults).
- This is a particularly visual lecture. Slide material on folds and faults can be confusing if you have never seen these geologic features before. It will help greatly if you preview Figure Story 11.6 before coming to lecture.
- It will also be useful to bookmark the key figures (Figure Story 11.6 and Figures 11.11 and 11.16) in your text for quick reference during lecture.

Note-Taking Tip

Figures of faults and folding can be drawn very simply...once you understand them. But until you do, make it easy on yourself. Photocopy Figure Story 11.6 and Figures 11.11 and 11.16. Three-hole punch them for easy insertion into your three-ring notebook. With them already in your notebook before lecture, you will not get distracted drawing them during lecture.

After Lecture

Education is a voyage in self-discovery.

—Laurence M. Gould

The perfect time to review your notes is right after lecture. The following checklist contains both general review tips and specific suggestions for this chapter.

Check your notes: Have you...

- ☐ added additional visual material? *Suggestion: Test your understanding of Chapter 11 by adding simple sketches of important geological features, such as normal, reverse, and thrust faults, plus an anticline, syncline, asymmetric fold, and overturned folds. Insert these sketches into your notes. Add captions to help you keep things straight. (For example, I associate the word "sink" to the "s" in syncline. "Sink" is what a syncline resembles.)*
- ☐ created a brief "big picture" overview of this lecture (using a sketch or written outline)? *Suggestion: Figure Story 11.6 is a good one-page summary of rock deformation.*

Intensive Study Session

Because there is a lot to learn in this chapter, be sure to set priorities for studying. Quite likely, there is more material than you will have time to study in one intensive study session. We recommend you give highest priority to activities that involve answering questions. Answering questions while using your text and lecture notes as reference material is far more efficient than reading chapters or glancing over notes. As always you have three sources from which to choose questions.

- **Practice Exercises and Review Questions.** Be sure to do Exercise 1. It gets to the key information you need to learn in this chapter.
- **Text.** This is a very visual chapter, so focus your attention on understanding the figures. Most of the really essential material for this chapter will be found in Figure Story 11.6 and Figures 1.11 and 1.16. Focus on understanding these illustrations first. Answer Exercises 1, 4, and 5 at the end of the chapter.
- **Web Site Online Review Exercises and Study Tools**
  http://www.whfreeman.com/understandingearth
- Complete the **Web Review Questions**. Pay particular attention to the hints if you get stumped. The **Geology in Practice** exercise: *Folds, Faults, and Other Records of Rock Deformation* provides an opportunity for you to practice your skills identifying and interpreting classic examples of rock deformation. The

Flash Cards will help you learn the new terminology. Check out the **Photo Gallery** for more striking images of rock deformation. The following **Online Review Exercises** will help you organize your knowledge about the different features of folds and faults: *Tectonic Forces in Rock Deformation Review* and *Identify the Type of Fold*.

TECTONIC STRESS LEADS TO STRAIN

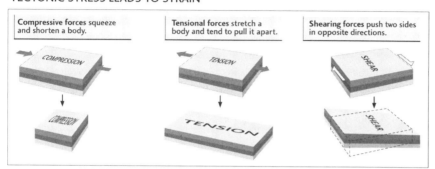

Exam Prep

Materials in this section are most useful during your preparation for midterm and final examinations. The following **Chapter Summary** and **Practice Exercises and Review Questions** should simplify your chapter review. Read the **Chapter Summary** to begin your session. It provides a helpful overview that should refresh your memory.

Next, work on the **Practice Exercises and Review Questions.** In order to determine how you stand on mastery of this chapter, complete the exercises and questions just as you would a midterm. After you answer the questions score them. Finally, review any question that you missed. Identify and correct the misconception(s) that resulted in your answering the question incorrectly.

Chapter Summary

How do rocks deform (bend and break)?

- All rocks may bend (ductile behavior) and break (brittle behavior) in response to the application of forces. Laboratory experiments have revealed that whether a rock exhibits ductile or brittle behavior depends on its composition, temperature, depth of burial (confining pressure), and rate with which tectonic processes apply force.
- Ductile behavior is more likely when a rock is exposed to higher temperatures, deeper burial, slower application of tectonic forces, and is a sedimentary rock. Brittle behavior is favored when rocks are cooler, closer to the Earth's surface, exposed to more rapid application of tectonic forces, and is an igneous or high-grade metamorphic rock.

What geologic features are produced when rocks deform?

- Folding is a result of ductile deformation. From the type of fold and its orientation geologists can interpret the orientation of the tectonic forces and characteristics of the rock layers during deformation.
- Faulting and jointing are a result of brittle deformation. Jointing occurs when a rock fractures but there is little movement along the fracture planes. Faults are fractures along which there is appreciable movement (offset). The type and orientation of faults and joints provides valuable information about the tectonic forces and the characteristics of the rock layers at the time of deformation.

What do geologic structures produced by rock deformation tell geologists about the geologic history of a region?

- The type of fold or fault provides a basis for geologists to interpret the type of tectonic force acting on the rock during deformation. Tectonic forces can be of three types: compressive, tensional, and shearing forces. These same kinds of forces are active at all three types of plate tectonic boundaries: compressive forces dominate at convergent boundaries (where plates collide or subduct); tensional forces dominate at divergent boundaries (where plates are pulled apart); and shearing forces dominate at transform faults (where plates slide horizontally past each other).
- Geologic structures, such as folds, faults, and joints, occur on all scales from microscopic to the size of a mountainside. Geologists deduce the geologic history of a region in part by unraveling the history of deformation, thereby reconstructing what the rock units looked like before deformation.
- Regional deformational fabrics can help geologists decipher the plate tectonic history for the region. Refer to Figures 11.21 and 11.22.

Practice Exercises and Review Questions

Answers and explanations are provided at the end of the Study Guide.

Exercise 1: Silly Putty®

Silly Putty® is a popular teaching aid with geologists because, at room temperature, it exhibits all three kinds of deformation characteristic of solids. If you pull on the putty quickly, it will snap into two pieces. It is easy to bend and mold the putty into many shapes. Plus, if you throw a ball of it on the floor, the ball will bounce. Compare the properties of Silly Putty® with the behavior of rocks by completing the table below.

Behavior of Silly Putty®	Behavior of rock	Type of force	Geologic structure produced by this style of deformation
Snaps into pieces		*Tensional*	
Bends	*Ductile*		
Bounces	*Elastic—Rocks do exhibit elastic behavior. More on this when we study earthquakes.*	*Compressional—The ball of putty is compressed by the impact with the floor.*	NOTE: *Earthquakes are attributed to the elastic properties of rocks.*

Exercise 2: Geologic Structures

For each of the following five illustrations of deformed rocks, name the (A) geologic structure, e.g., normal fault, syncline; (B) type of force, e.g., compressional, tensional, shearing force, responsible for producing each geologic structure; and (C) the plate tectonic boundary, e.g., convergent, divergent, or shear, with which the geologic structure is commonly associated.

A. Geo structure _____   VERTICAL SECTION
B. Type of force _____
C. Commonly associated plate tectonic boundary
   _____

D. Geo structure _____   VERTICAL SECTION
E. Type of force _____
F. Commonly associated plate tectonic boundary
   _____

G. Geo structure _____   VERTICAL SECTION
H. Type of force _____
I. Commonly associated plate tectonic boundary
   _____

J. Geo structure _____   OBLIQUE VIEW
K. Type of force _____
L. Commonly associated plate tectonic boundary
   _____

M. Geo structure _____   VERTICAL SECTION
N. Type of force _____
O. Commonly associated plate tectonic boundary
   _____

Exercise 3: Anticline vs. Syncline

A. Briefly describe the diagnostic differences between an anticline and a syncline.

B. Then, draw a picture of a typical outcrop pattern for a plunging syncline exposed at the surface. Refer to Figure 10.17.

A. _____   _____
_____
_____
_____

B. Draw an outcrop pattern for a plunging syncline.

Exercise 4: Identifying Geologic Structures

Fill in the blanks with the correct name for the geologic structure illustrated.

A. This fold is called a/an _____.

B. This fold is called a/an _____.

C. This fault is called a/an _____.

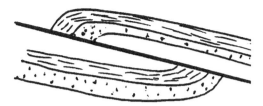

D. This fault is called a/an _____.

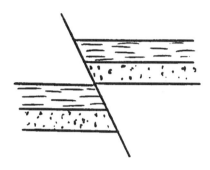

E. This fault is called a/an _____.

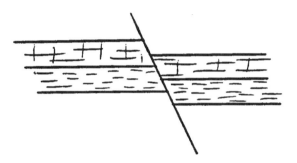

Review Questions

Answers and explanations are provided at the end of the Study Guide.

1. A rock that breaks suddenly in response to the application of forces is
   A. elastic.
   B. plastic.
   C. brittle.
   D. ductile.

2. The two measurements that define the orientation of an exposed rock layer are
   A. strike and dip.
   B. strike and slip.
   C. slip and dip.
   D. fold axis and tilt.

3. The sense of motion along the San Andreas fault in California is
   A. left-lateral strike slip.
   B. right-lateral strike slip.
   C. dip slip.
   D. thrust.

4. As tensional forces are applied to a continental region, the resulting geologic feature will be a/an
   A. anticline.
   B. rift valley.
   C. thrust fault.
   D. dome.
   Hint: Refer to Figure Story 11.6.

5. When no offset can be detected along a fracture in a rock, the fracture is called a
   A. stress plane.
   B. joint.
   C. fault.
   D. rupture.

6. Of the following conditions, which one promotes ductile deformation of rocks?
   A. Old igneous rocks within the interior of a continent
   B. Rocks deforming at a relatively low temperature
   C. Stress building up rapidly to a very high level
   D. Rocks subjected to high confining pressures and temperatures

7. When deformed, which of the following rocks is more likely to fracture brittlely instead of flow ductilely?
   A. Basalt
   B. Shale
   C. Pure limestone
   D. Muddy sandstone

8. As molten rock cools near or at the surface, it can develop shrinkage fractures that extend vertically down into the rock body. These crisscrossing, regularly patterned fractures create long thin rods of rock we call
   A. shrinkage palisades.
   B. tension faults.
   C. columnar joints.
   D. elongate joints.
   Hint: Refer to the photo at the beginning of Chapter 5.

9. Which of the following geologic structures is caused by tensional forces?
   A. Thrust fault
   B. Reverse fault
   C. Anticline
   D. Normal fault

10. Thrust faults commonly form
    A. around hot spots.
    B. where continents are colliding.
    C. where continents are pulling apart.
    D. along a transform fault.

11. Which of the following defines the direction of dip?
    A. A line at right angles to the strike line
    B. A line north of the strike line
    C. A line parallel to the strike line
    D. A line parallel to plunge
    Hint: Refer to Figure 11.4 in your textbook.

12. From an airplane you notice that the outcrops of tilted layers of rock make a distinct zigzag pattern across a plain. You reasonably conclude that the
    A. area has been cut by numerous normal faults.
    B. layers are folded into a series of plunging folds.
    C. layers are folded into a series of nonplunging folds.
    D. layers have been tilted so that all the layers dip in the same direction.
    Hint: Refer to Figures 11.17 and 11.18 in your textbook.

CHAPTER 12

Mass Wasting

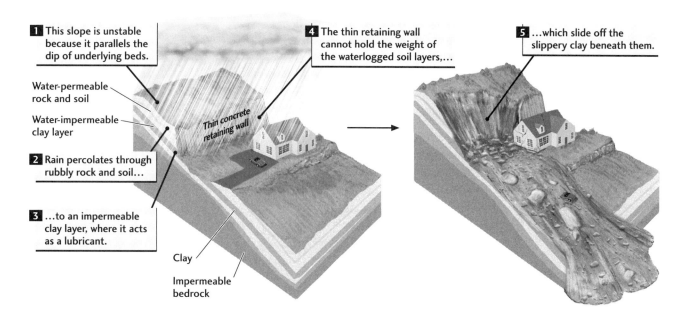

Earth Policy Box 12.1. This slope is unstable because it parallels the dip of the underlying rock layers and rests on a clay layer that would act as a lubricant if it became waterlogged. The slope in the back of the house has been oversteepened, and the concrete retaining wall is too thin to hold it.

Before Lecture

Chapter Preview

- **What is mass wasting?**
  Brief answer: Mass wasting, also called mass movement, is the down-slope movement of rock material.
- **Why do mass movements occur?**
  Brief answer: The three most important factors enhancing the potential for mass movements are: the steepness of the slope; the nature of the rock making up the slope; and the water content.
- **How can damage from mass movements be minimized?**
  Brief answer: Careful engineering and restricting land use can minimize the hazards associated with mass movements.

Previewing Tip

It will be very helpful to work on Exercise 1 before going to lecture. Complete the exercise and take it to class with you. Your lecturer will probably show slides to help you understand the different kinds of mass wasting. You will understand these differences better if you have worked on them before class.

Vital Information from Other Chapters

The composition and internal fabric of rocks significantly influence the rock's strength and the potential for mass movement. Therefore, a review of the composition and especially of the different kinds of fabrics (textures) for igneous, sedimentary, and metamorphic rocks will provide you with vital information for understanding the different circumstances that cause mass movements.

While reviewing the basic rock textures described in Chapters 4, 5, 6, 8, and 9 ask yourself what textures might contribute to a weaker rock and an enhanced potential for mass movement.

During Lecture

One goal for lecture should be to leave class with a good set of answers to the preview questions.

- To avoid getting lost in details, keep the "big picture" in mind. Chapter 12 tells the story of what causes mass wasting. Just remember, mass wasting is about classification and your job is to understand the differences between the types of wasting. Be sure to pay close attention to comments your instructor may make about how each kind of mass wasting *differs* from the others. **Hint:** In general these differences will be about the *steepness* of the slope, *kind of rock* in the slope being moved, and the *water content*.
- You may not be familiar with the kinds of mass wasting (rock avalanche, creep, earthflows, etc.). To help you visualize this process, your lecturer may show slides of various kinds of mass wasting. Some of these may be very dramatic (e.g., landslides). Enjoy the drama and excitement!
- If you completed Exercise 1 before class, you can refer to it as your lecturer talks about the different kinds of mass wasting.

After Lecture

The perfect time to review your notes is right after lecture. The following checklist contains both general review tips and specific suggestions for this chapter.

Check your notes: Have you...

- ☐ added simple sketches to your notes? This will help you remember the key aspect of each kind of mass wasting. Hint: Your sketch need not be artistic to be useful. Sketch only the features you need to remember. Example: For a rock avalanche you could draw a steep slope (one line at a 45 degree angle) with a pile of large blocks at the bottom to designate large masses of broken rock (see Figure 12.9).
- ☐ added a comparison chart to your notes that will help you master the classification of different kinds of mass wasting? **Hint:** See Exercise 1 in the **Practice Exercises and Review Questions.**
- ☐ created a brief "big picture" overview of this lecture (using a sketch or written outline)? *Suggestion for Chapter 12: Write a brief summary of the most important points you have learned from this chapter that might influence your choice of future home sites.*

Intensive Study Session

Set priorities for studying this chapter. There is a lot to do, quite likely more than you will have time for in one intensive study session. Remember, you should always look to your instructor first when you are attempting to set your studying priorities. Pay particular attention to any exercises recommended by your instructor during lecture and *always* answer those first. Your instructor is also your best resource if you are wondering which material is most important. In addition, we recommend you give highest priority to activities that involve answering questions. Answering questions while using your text and lecture notes as reference material is far more efficient than reading chapters or glancing over notes. As always you have three sources from which to choose questions.

- **Practice Exercises and Review Questions.** Next, use the **Practice Exercises and Review Questions.** Be sure to do Exercise 1. It gets to the key information you need to learn in this chapter.
- **Text.** Work on your responses to Exercises 3 and 5 and Thought Questions 1, 3, and 6 at the end of Chapter 12 in the textbook. **Hint:** Table 12.1 and Figure 12.6 will help you sort out the distinguishing features of the different kinds of mass movement.
- **Web Site Online Review Exercises and Study Tools**
  http://www.whfreeman.com/understandingearth
  Complete the **Concept Self-Checker** and **Web Review Questions** located on the web site. Pay particular attention to the hints and explanations for the answers. Also at the web site are **Flash Cards** to help you learn new terms. The **Online Review Exercises:** *Identify the Rock Mass Movement* and *Identify the Unconsolidated Mass Movement,* will help you review the names for different kinds of mass movement. After completing these short exercises, be sure to compare the illustrations with photos of different kinds of mass movement that appear in your textbook and in the online **Photo Gallery.** As you study the images, assess what factors were most important in causing slope instability. Remember that the three most important factors enhancing the potential for mass movements are the steepness of the slope, the nature of the rock material in the slope, and the water content.

Exam Prep

Materials in this section are most useful during your preparation for midterm and final examinations. The following **Chapter Summary** and **Practice Exercises and Review Questions** should simplify your chapter review. Read the **Chapter Summary** to begin your session. It provides a helpful overview that should refresh your memory.

Next, work on the **Practice Exercises and Review Questions.** In order to determine how you stand on mastery of this chapter, complete the exercises and questions just as you would a midterm. After you answer the questions score them. Finally, review any question that you missed. Identify and correct the misconception(s) that resulted in your answering the question incorrectly.

Chapter Summary

What is mass wasting?

- Mass movements (also called mass wasting) are slides, flows, or falls of large masses of rock material down slopes when the pull of gravity exceeds the strength of the slope materials. Such movements can be triggered by earthquakes, absorption of large quantities of water from torrential rainfall, undercutting by flooding rivers, human activities, or other geologic processes.

Why do mass movements occur?

- The three most important factors enhancing the potential for mass movements are: the steepness of the slope; the nature of the rock making up the slope; and the water content. Although steep slopes are prone to mass movements, slopes of only a few degrees can also fail catastrophically because of these other factors.
- Slopes become unstable when they become steeper than the angle of repose (the maximum slope angle that unconsolidated material will assume). Slopes in consolidated material may also become unstable when they are oversteepened or denuded of vegetation. Erosion by rivers and glaciers and human activities can oversteepen slopes and, thereby, increase the potential for mass movement.
- The composition, texture, and geologic structure of the slope material is another important factor influencing the potential for slope failure. For example, rocks with high clay content tend to be weak and may liquefy. Tilted layers of sedimentary or volcanic rocks are more likely to fail along bedding planes when the bedding parallels the slope. Failure of foliated metamorphic rocks is more likely to occur parallel to the direction of foliation.
- Water absorbed by the slope material contributes to instability in two ways: (1) by lowering internal friction (and thus resistance to flow) and (2) by lubricating planes of weakness in the slope.

How can damage from mass movements be minimized?

- The hazards and damage associated with mass movements can be minimized by careful geological assessment, engineering, and land use policies that restrict development on unstable slopes. Of particular importance is the avoidance of steepening or undercutting slopes and minimizing the amount of water that can infiltrate the slope material. In some areas particularly prone to mass movements, development may have to be restricted.

Practice Exercises and Review Questions

Answers and explanations are provided at the end of the Study Guide.

Exercise 1: Inventory of the Different Kinds of Mass Wasting

The authors discuss eight different kinds of mass wasting. As an aid to learning the circumstances that favor each of these types of mass movement, use your textbook to fill in the blanks in the table below. Textbook figures, figure captions, and photographs will help you complete the table.

Hint: You probably haven't seen many of these features before, so be sure to examine the photos and figures of each type of mass wasting in your textbook or on the online photo gallery. If you are a visual learner, this activity may be vital. Also, to get a kinesthetic feel for these movements imagine yourself trying to outrun each movement. Indicate in the space labeled "Speed" whether you could escape the mass movement by walking, running, or moving as fast as a speeding auto.

Kind of mass wasting	Composition of slope (consolidated vs. unconsolidated and wet vs. dry)	Characteristics
Rock avalanche		Speed: running or a speeding auto Slope angle: steep slopes Triggering event(s): earthquakes Notes: Occur in mountainous regions where rock is weakened by weathering, structural deformation, weak bedding, or cleavage planes
Creep		Speed: Slope angle: any angle Triggering event(s): none Notes:
Earthflows		Speed: Slope angle: any angle Triggering event(s): intense rainfall Notes: Fluid-like movement
Debris flow		Speed: Slope angle: any angle Triggering event(s): Notes:
	Mostly finer rock materials with some coarser rock debris with large amounts of water	Speed: Slope angle: Triggering event(s): intense rainfall or catastrophic melting of ice and snow by a volcanic eruption. Notes: Contains large amounts of water
Debris avalanche	Water-saturated soil and rock	Speed: Slope angle: Triggering event(s): Notes:
Slump		Speed: walking or running Slope angle: any slope Triggering event(s): rainfall Notes:
	Surface layers of soil	Speed: walking Slope angle: any angle Triggering event(s): Notes: Occurs only in cold regions when water in the surface layers of the soil alternately freezes and thaws. Water cannot seep into the ground because deeper layers are frozen.

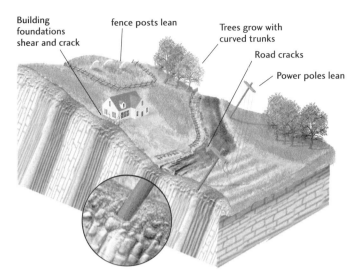

Figure 12.10. Mass movement is not restricted only to steep slopes. Soil creep (shown here), solifluction, earthflows, mudflows, and the liquefaction of clay layers can occur in nearly horizontal layers (also refer to the photograph in Figure 12.5).

Exercise 2: Water's Role in Mass Wasting

Water enhances the potential for mass wasting in many ways. Using your textbook as a guide, briefly describe five different ways water enhances the potential for mass movements.

1. _____
2. _____
3. _____
4. _____
5. _____

Exercise 3: Evaluation of Slope Stability

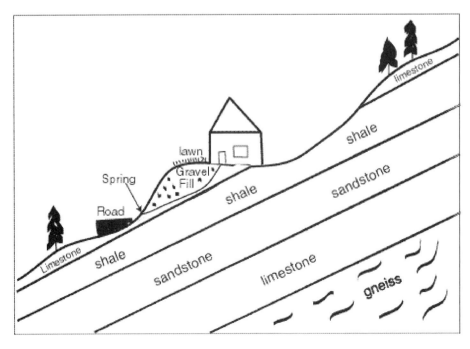

A. Discuss three factors that enhance the potential for mass movement at the home site shown on the previous page.

1. _____

2. _____

3. _____

B. Given that the home is already built on this site, briefly discuss two possible ways of reducing the risk of damage to the house due to slope failure.

1. _____

2. _____

Test-Taking Tip

Time permitting, it is sometimes helpful to sketch the alternatives to a test question. For example, if you aren't sure about the answer to Review Question 6, then you could sketch "rock layers at right angles to the slope," "rock layers parallel to the slope," etc. Sketching can be particularly useful to kinesthetic learners, as the action of writing may "jog" or "unlock" your memory when you are stuck.

Review Questions

Answers and explanations are provided at the end of the Study Guide.

1. Which hillside home site is the best long-term investment?

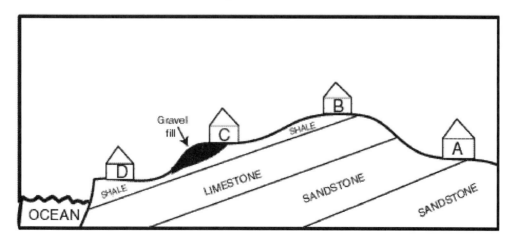

A. Site A
B. Site B
C. Site C
D. Site D

2. What force drives mass wasting?
   A. Heat
   B. Gravity
   C. Friction
   D. Convection

3. Mass wasting tends to occur when
   A. a slope becomes steeper due to undercutting by a river or ocean waves.
   B. the mass on a slope decreases by draining water from the ground.
   C. friction is increased by draining water from the ground.
   D. friction is decreased by taking water out of the ground.

4. Talus consists largely of
   A. clay and other very fine rock particles.
   B. a mixture of powdered rock and ice.
   C. coarse, angular rock fragments.
   D. alternate layers of sand, silt and clay.
   Hint: See the *Talus slope near Hofn, Iceland* in the **Photo Gallery** on the web site.

5. The angle of repose is
   A. that angle at which rock material is most stable.
   B. that angle at which lava flows will solidify without spreading out.
   C. the angle of a slope that will no longer support large boulders and rock pillars.
   D. the maximum slope at which a slope of loose material will lie without cascading down.

6. An important factor in mass wasting is the orientation of rock layers, foliation, or jointing. For layered sedimentary and volcanic rocks, which condition is the least stable?
   A. Rock layers are at right angles to the slope.
   B. Rock layers are parallel to the slope.
   C. Rock layers are horizontal to the slope.
   D. Rock layers stand vertical.

7. Which of the following would be most subject to mass movements (assume the slope and climate is the same in each case)?
   A. High-grade gneiss, with highly contorted foliation
   B. Quartz-cemented sandstone, with layering perpendicular to the slope
   C. Shale, with bedding dipping parallel to the slope
   D. Massive granite bedrock

8. Solifluction usually occurs in
   A. cold regions.
   B. very cold regions, such as Antarctica.
   C. any area where there is lots of sunshine.
   D. tropical regions.

9. What is the most effective way to stabilize an active landslide, given the options below?
   A. Piling additional rock and soil material on the landslide near its top
   B. Saturating the landslide itself with water
   C. Draining the water away from and out of the landslide area
   D. Cutting away the toe (base) of the landslide

10. Roads through mountainous regions tend to be unstable and require more maintenance if they are built on
    A. bedrock, such as granite.
    B. horizontal lava flows.
    C. rock layers that dip perpendicular to the hill slope.
    D. rock layers that dip parallel to the hill slope.

11. Your beautifully landscaped house, built on an idyllic Georgia hillside setting of small, irregularly undulating knolls and depressions, with trees tilted at interesting angles, has developed a bad case of broken and shifting foundation. The probable cause for the foundation problem is

    A. melting permafrost.
    B. that the house is built on an active earthflow.
    C. mudflow from an active nearby volcano.
    D. root wedging from the trees.

12. Homeowners in California who survived recent wildfires are not quite "out of the woods" yet. With the approaching rainy season their next problem will be

    A. increased potential for mudflows and debris flows.
    B. accelerated soil erosion.
    C. flash floods.
    D. all of the above.

CHAPTER 13

The Hydrologic Cycle and Groundwater

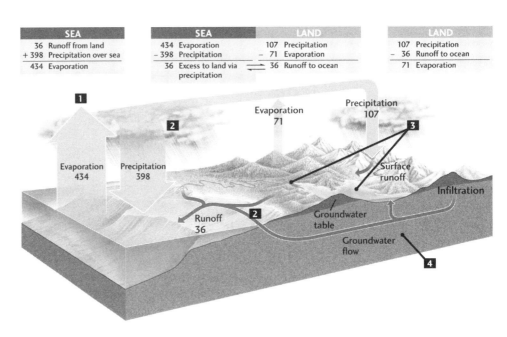

Figure 13.2. The hydrologic cycle.

Before Lecture

Before you attend the lecture be sure to spend some time previewing Chapter 13. For an efficient preview, use the questions below. **Chapter Preview** questions constitute the basic framework for understanding this chapter. Preview works best if you do it just before lecture. With the main points in mind you will understand the lecture better. This in turn will result in a better and more complete set of notes.

How much time should you devote to preview? Obviously, more time is better than less. But even a brief (5- or 10-minute) preview session just before lecture begins will produce a result you will notice. For a refresher on why previewing is so important, see **How to Be Successful in Geology** in Chapter 3 (Part I) of the Study Guide.

Chapter Preview

- **How does water move in and around the Earth?**
  Brief answer: The hydrologic cycle is a model for the movement of water on Earth. Refer to Figure 13.2.
- **How does water move below the ground surface?**
  Brief answer: Porosity and permeability are the principle factors that control the infiltration and flow of groundwater (skim the section titled *How water flows*).
- **What factors govern our use of groundwater resources?**
  Brief answer: Water like soil and air is one of our most basic natural resources. Some of the most important factors governing our ability to use groundwater are the depth to the groundwater table, springs and artesian systems, the balance between recharge and discharge, Darcy's Law, and water quality.
- **What geologic processes and features are associated with groundwater?**
  Brief answer: Groundwater is the geologic agent responsible for caves, karst topography, and the formations that decorate caves. Groundwater is a vital component of geothermal systems. See Figures 13.19, 13.20, and 13.22.

During Lecture

Warm-Up Activity

This is a very interesting chapter! Use some of the interesting material to warm you up for lecture. Spend 5 to 10 minutes just before lecture browsing *Earth Issues* 13.1 "Water Is a Precious Resource: Who Should Get It?" and *Earth Issues* 13.2 "When Do Groundwaters Become Nonrenewable Resources?" Which issue interests you the most? After browsing ask yourself what you would most like to learn from this chapter and lecture.

One goal for lecture should be to leave class with a good set of answers to the preview questions.

- To avoid getting lost in details, keep the "big picture" in mind. Chapter 13 is a survey of water in and around the Earth. It tells the story of the hydrologic cycle: how water moves in and around the Earth and maintains a balanced water flow budget.
- First, focus on understanding Figure 13.2 (The hydrologic cycle).
- Then, work on understanding the factors that govern the flow of groundwater, illustrated in the figures on pages 285–289.

After Lecture

The perfect time to review your notes is right after lecture. The checklist below contains both general review tips and specific suggestions for this chapter.

Check your notes: Have you...

- ☐ included a clear representation of the hydrologic cycle somewhere in your notes (see Figure 13.9)?
- ☐ clearly identified factors that govern our use of groundwater; e.g., depth of the groundwater table, springs and artesian systems, the balance between recharge and discharge, and Darcy's Law?
- ☐ added visual material to help you understand the material? *Suggestions for Chapter 13. Sketch a simple version of Figure 13.7 (porosity of rock materials). Photocopy a copy of Exercise 1 after you have completed the chart and insert it into your notes. This will be great aid for exam review: it summarizes all you will need to know about porosity.*
- ☐ created a brief "big picture" overview of this lecture (using a sketch or written outline)? *The hydrologic cycle (Figure 13.2) provides a good visual summary of the chapter. Activate kinesthetic learning by adding a simplified version to your notes.*

Intensive Study Session

Set priorities for studying this chapter. We recommend you give highest priority to activities that involve answering questions. We recommend the following strategy for learning this chapter.

- **Add illustrations to your lecture notes.** First, preview the key figures in Chapter 13. They are Figures 13.2, 13.7, 13.9, 13.10, 13.11, 13.12, 13.14, 13.15, and 13.20. Insert simple sketches of these figures into your lecture notes.
- **Practice Exercises and Review Questions.** Next, complete Exercises 1 and 2. You will get the greatest return on your study time by working on these exercises because they will help you remember important ideas in the chapter. Then, work on answering each of the review questions to check your understanding of the lecture. Check your answers as you go, but do try to answer the question before you look at the answer. Pay attention to the test-taking tips we provide. They will help you do better on quizzes and exams.
- **Text.** Before the next exam, complete all 11 exercises at the end of Chapter 13 in the text. These require short answers and won't take long if you know the material. Note that helpful animations are provided on the web site for Exercises 3, 5, and 6.
- **Web Site Study Resources**
  http://www.whfreeman.com/understandingearth
  Complete the **Web Review Questions.** Pay particular attention to the explanations for the answers. Be sure to check out the animations of an aquifer, confined aquifer, and the dynamic balance between recharge and discharge. The *Geology in Practice* exercises allow you to participate in an ongoing case study involving ground contamination.

The frog does not drink up the pond in which it lives.

—NATIVE AMERICAN PROVERB

Exam Prep

Materials in this section are most useful during your preparation for midterm and final examinations. The following **Chapter Summary** and **Practice Exercises and Review Questions** should simplify your chapter review. Read the **Chapter Summary** to begin your session. It provides a helpful overview that should refresh your memory.

Next, work on the **Practice Exercises and Review Questions.** In order to determine how you stand on mastery of this chapter, complete the exercises and questions just as you would a midterm. After you answer the questions score them. Finally, review any question that you missed. Identify and correct the misconception(s) that resulted in your answering the question incorrectly.

Chapter Summary

How does water move in and around the Earth?

- The hydrologic cycle is a flowchart or model for the distribution and movements of water on and below the surface of the Earth. The major reservoirs for the hydrologic cycle are oceans, glaciers, groundwater, lakes and rivers, the atmosphere, and biosphere in decreasing volumes. Water moves in and out of these reservoirs by various pathways and at varying rates. Over the short term a balance is maintained among the major reservoirs at and near the Earth's surface. However, climate change and longer-term tectonic processes, such as mountain building and human activity, can alter the rate of water movement between reservoirs and impact the size of the reservoirs.

How does water move below the ground surface?

- The infiltration of water into the ground and groundwater flow are largely controlled by the porosity and permeability of the rock materials and topography. A groundwater aquifer is in dynamic balance between recharge (the amount of water that infiltrates into the aquifer) and discharge which can occur from springs or wells.

What factors govern our use of groundwater resources?

- Darcy's Law describes the groundwater flow rate in relation to the slope of the water table and the permeability of the aquifer.
- Human demand for groundwater has increased to a level where pumping discharges from many aquifers exceeds the natural rates of recharge. As a result, aquifers are being depleted and groundwater tables are lowering to a point where dependable, quality groundwater is becoming more and more of a challenge to supply.
- Water quality may be compromised by both natural and human sources of contamination. Various factors, such as recharge rate and aquifer size, influence the amount of effort and effectiveness of attempts to clean up contamination.

What geologic processes and features are associated with groundwater?

- Caves, sinkholes, and associated karst topography are a result of the dissolution of carbonate rocks (limestone) by groundwater. Karst topography is well developed in regions of high rainfall, abundant vegetation, underlying fractured limestone, and an appreciable hydrologic gradient to enhance groundwater flow rates. Environmental problems associated with karst regions include surface subsidence from collapse of underground space and catastrophic cave-ins and sinkhole formation.
- All rocks below the groundwater table are saturated with water. With increasing depth, porosity and permeability typically decrease as confining pressure increases. Water temperature increases progressively with increasing depth and, as a result, the water dissolves more solids. Hot springs and geysers are surface expressions of the circulation of hydrothermal waters over a magma body or along a deep-seated fault.

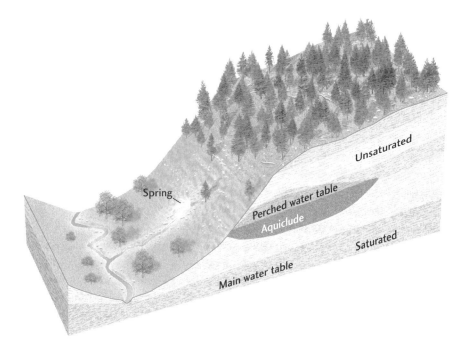

Figure 13.11. A perched water table forms in geologically complex situations.

Practice Exercises and Review Questions

Answers and explanations are provided at the end of the Study Guide.

Exercise 1: Evaluating Rock Materials as Potential Aquifers

You have recently purchased a rustic country cabin and need to drill a new well for a dependable water supply for the cabin. The geology around your cabin is complex due to ancient mountain building events. Because the rocks are folded and faulted, it is difficult to predict what rock might be encountered as the water well is drilled. Which of the follow rock materials has the potential of yielding groundwater to your well? Fill in the blank parts of the table below.

Hint: Keep in mind that generally permeability increases as porosity increases, but not always. Permeability also depends on the sizes of the pores, how well they are connected, and how tortuous a path the water must travel to pass through the material. Refer to Figure 13.7 and Table 13.2.

Rock material	Porosity (high, medium, low)	Potential as an aquifer (good, moderate, poor)
Loose, well-sorted, coarse sand		
Silt and clay	Low	
Granite and gneiss		Poor
Highly fractured granite		
Sandstone	Medium	
Shale		
Highly jointed limestone		Moderate to good

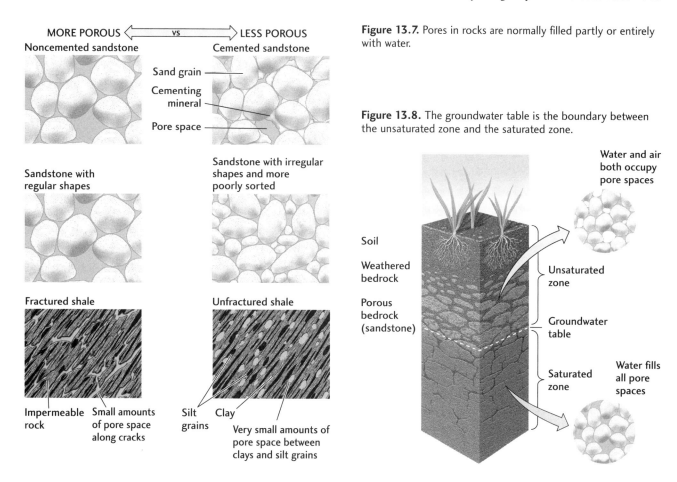

Figure 13.7. Pores in rocks are normally filled partly or entirely with water.

Figure 13.8. The groundwater table is the boundary between the unsaturated zone and the saturated zone.

Exercise 2: Evaluating Groundwater Wells

Fill in the blanks below with either "high," "low," or "none" for your evaluation of the potential characteristic of each well.

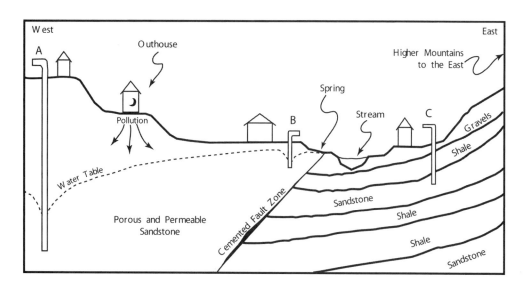

Well A—potential for: 1. pollution _____

                           2. artesian flow _____

                           3. discharge _____

Well B—potential for: 4. pollution _____

5. artesian flow _____

6. discharge _____

7. long-term supply _____

Well C—potential for: 8. pollution _____

9. artesian flow _____

10. discharge _____

Americans now drink more soda pop than water from the kitchen tap—47 gallons of soda pop to only 37 gallons of water per person each year.

—WORLD WATCH, 1990.

Review Questions

Answers and explanations are provided at the end of this Study Guide.

1. In the hydrologic cycle, how does the evaporation rate from the land surface compare to the evaporation rate off the oceans?
   A. The evaporation rate from the land is much greater than from the oceans.
   B. The evaporation rate from the ocean is much greater than from the land.
   C. Evaporation rates from the land and oceans are equal.
   D. There is no reasonable comparison because it is too wet over the oceans for evaporation to occur.
   Hint: Refer to Figure 13.2.

2. The oceans contain by far the most amount of water on Earth. What is the second largest reservoir for water on Earth?
   A. Lakes   C. Rivers
   B. Groundwater   D. Polar ice and glaciers

3. The oceans contain about how much of the water in the hydrosphere?
   A. 96%   C. 50%
   B. 80%   D. 35%

4. What happens to the porosity as the grain size gets smaller?
   A. It increases.   C. It decreases.
   B. It remains unchanged. D. None of the above.

5. The water table is
   A. the top of the unsaturated zone.
   B. the top of the saturated zone.
   C. generally present only in moist climates.
   D. the contact between an aquifer and underlying, impermeable layer of rock.

6. The ability of a solid, such as rock, to allow fluids to pass through it is
   A. discharge.   C. permeability.
   B. capillary fringe.   D. porosity.

7. An icicle-like deposit hanging from the ceiling of a cave is a
   A. stalactite.   C. stalagmite.
   B. karst formation.   D. quartzite.

8. Of the following rock types, which is the most susceptible to groundwater solution, therefore making it a formation most likely to have caves?
   A. Granite   C. Limestone
   B. Sandstone   D. Shale

9. A rock or soil layer that is water-bearing is
   A. a perched water table.
   B. a zone of aeration.
   C. a stratum.
   D. an aquifer.

10. The potential for geothermal energy is highest in a region that has numerous
    A. surface lakes.
    B. caves.
    C. hot springs.
    D. sinkholes.

11. An aquiclude is
    A. a confined aquifer.
    B. always located at the top of the water table.
    C. a rock layer that provides a good flow of water into a well.
    D. an impermeable rock layer that does not allow water to flow through it.

12. Which of the following would make the best aquifer?

	Porosity	Permeability
A. Rock A	5%	high
B. Rock B	10%	medium
C. Rock C	30%	low
D. Rock D	35%	medium

13. If all other conditions are equal, groundwater will move faster
    A. where sand grains are very well cemented.
    B. through loose sand than through clay.
    C. where permeability of the aquifer is lower.
    D. through clay than through sand.
    Hint: Refer to Figure 13.7.

14. A perched water table will most likely develop on top of
    A. shale.
    B. highly fractured granite.
    C. gravel.
    D. sandstone.
    Hint: Refer to Figure 13.11.

15. Rivers and streams that flow all year long, even during long periods without rain, are likely to be fed by
    A. sinkholes.
    B. springs.
    C. wells.
    D. karst conditions.
    Hint: Refer to Figure 13.11.

Test-Taking Tip

When you are unsure about an answer sometimes you can make a correct guess just by looking at the responses. Long responses are more likely to be correct than short ones. This occurs because when people write test items they have to be sure the correct alternative is clear and accurate and that may sometimes take more words. Obviously, this is a strategy to use only as a last resort, when you have no clue what the correct answer is. The safer strategy is to learn the material thoroughly.

16. An artesian well will flow if the
    A. top of the well is lower than the water table in the recharge area.
    B. top of the well is higher than the water table in the recharge area.
    C. bottom of the well is lower than the land surface in the recharge area.
    D. bottom of the well is lower than the water table in the recharge area.
    Hint: Refer to Figure 13.10.

17. Which well will exhibit artesian flow?

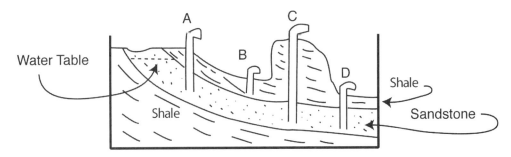

A. Well A
B. Well B
C. Well C
D. Well D

Hint: Refer to Figure 13.10.

Test-Taking Tip: Leveraging Correct Answers

Be alert for items that test the same idea or fact. Often you can use an item you are sure about to help you obtain a correct response for another item where you are not so sure. Example: Items 5 and 6 both test the same concept, namely what makes an artesian well flow. Let's say the picture helped you figure out item 6 but you had left item 5 blank or weren't sure of your answer. Now, after working item 6 you should be able to go back and answer item 5 correctly. In essence, you are learning as you take the test. Learning by answering questions is one of the best ways to master material. Sound familiar?

18. Which materials would make the best aquifer?
    A. Clay and silt
    B. Gravel and sand
    C. Unfractured granite
    D. Highly cemented sandstone

19. At a shallow depth, a well will most likely encounter a good water supply in which of the following locations?
    A. In granite on a ridge top
    B. In sandstone on a ridge top
    C. In shale in a valley bottom
    D. In sandstone in a valley bottom

    Hint: Make a sketch illustrating each situation.

Test-Taking Tip: Eliminating Incorrect Answers

Unsure which is the right answer? Try going down the alternatives one at a time. Cross out each answer that you think is incorrect. Hopefully, there will only be one alternative left. It will probably be the correct answer. Example: In the item above did you eliminate the two items that mentioned "ridge top"? If so you narrowed your choice to shale and sandstone. If you completed the practice exercises, the choice between shale and sandstone was an easy one. If not and this were a real test you would have to guess. But because you eliminated two of the incorrect items your odds of guessing correctly would have doubled!

20. Which well is most likely pumping polluted water?

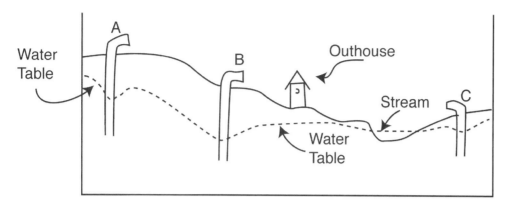

   A. Well A
   B. Well B
   C. Well C
   D. None

21. Of the wells illustrated in the diagram below, which one will produce the greatest water over the longest time?

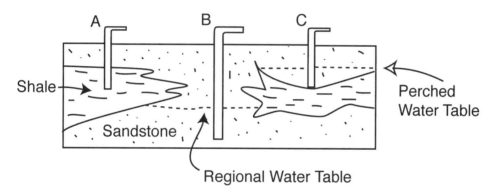

   A. Well A
   B. Well B
   C. Well C
   D. All of the wells will have high productivity.

22. If water is pumped from well K faster than natural recharge replenishes it, the result will most likely be

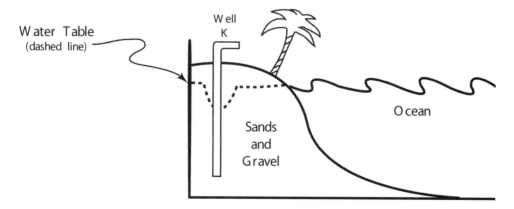

   A. sinkholes.
   B. ground subsidence.
   C. salt water intrusion.
   D. a dry well.
   Hint: Refer to Figure 13.14.

23. Which water source illustrated below is LEAST likely to be polluted?

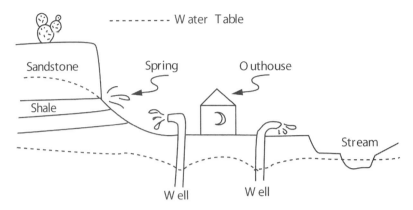

A. The spring
B. Well B
C. Well C
D. The stream

24. Assume that you are dealing with the same aquifer at four different localities and that A (cross-sectional area through which the water flows) and K (the hydraulic conductivity) are the same for each site. Given the following data on vertical drop (h) and flow distance (l) which well will produce water at the highest rate (Q)? Refer to Figure 13.15.

A. h = 20 meters and l = 500 meters
B. h = 30 meters and l = 1 kilometer
C. h = 300 meters and l = 6 kilometers
D. h = 600 meters and l = 100 kilometers

Figure 13.15. Darcy's law describes the rate of groundwater flow down a slope between two points, Elevation A and Elevation B. The volume of water flowing at a certain time (Q) is proportional to the difference in height (h) between the high and low points of the slope (here shown as the drop in elevation of the water table between the two points), divided by the flow distance between them (the hydraulic gradient, l) and by K, a constant proportional to permeability of the aquifer. The symbol A represents the cross-sectional area through which the water flows.

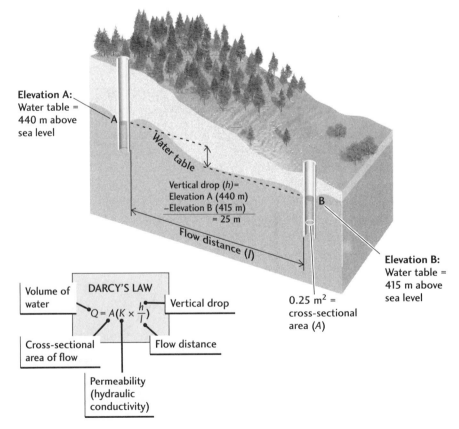

CHAPTER 14

Streams: Transport to the Oceans

I do not know much about gods, but I think that the river is a strong brown god—sullen, untamed, and intractable.

—T. S. Elliot, *The Dry Salvages*

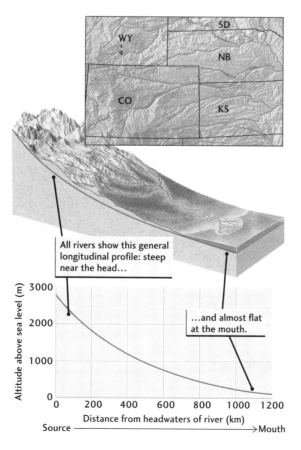

Figure 14.13. The longitudinal profile of the Platte and South Platte rivers from the headwaters of the South Platte in central Colorado to the mouth of the Platte at the Missouri River in Nebraska.

Before Lecture

As you preview this chapter keep in mind that the overarching question this chapter addresses is "How do streams work." Consider what the authors say:

"In this chapter, we focus on how streams accomplish their geological work: how water flows in currents; how currents carry sediment; how streams break up and erode solid rock; and, on a larger scale, how streams carve valleys and assume a variety of forms as they channel water downstream."

Understanding Earth

Chapter Preview

- **How does flowing water in streams erode solid rock and transport and deposit sediments?**
  Brief answer: Flowing water erodes rock by physical and chemical weathering processes. Turbulent flow is responsible for transporting sediments. When a stream flow slows, it loses its competence to carry sediment and deposits it.

- **How do stream valleys and their channels and floodplains evolve?**
  Brief answer: As a stream flows, it carves a valley. If the valley is wide enough the channel will be surrounded by a floodplain into which the channel will overflow during times of high water, dropping its sediment load as the water spreads out and slows. Channels may be straight, meandering (winding), or braided (divided, interlacing channels).

- **Hoes does a stream's longitudinal profile represent the equilibrium between erosion and sedimentation?**
  Brief answer: A stream's gradient (slope or longitudinal profile), base level, velocity (speed), discharge (amount of water), and the availability of sediment (load) determines whether a stream erodes, deposits, or reaches a balance between erosion and deposition—an equilibrium, called a graded stream. Changes in these stream characteristics can be used to predict the behavior of a stream when it is dammed or there is some other change in the drainage system.

- **How do drainage networks work as collection systems and deltas as distribution systems for water and sediment?**
  Brief answer: Rivers and tributaries constitute an upstream branching drainage network that collects the runoff for a particular drainage basin. Two notable examples are the Mississippi Basin and the Colorado River Basin. The nature of a drainage pattern depends on the basin topography, rock type, and geologic structure. See Figure 14.20. Deltas form at the mouths of rivers as the river branches into numerous distributaries and drops (distributes) its load of sediment. See Figures 14.20, 14.23, and 14.24.

Vital Information from Other Chapters

Chapters 13 and 14 are a package. Be sure to review material in the first half of Chapter 13. Pay particular attention to the role that streams play in the hydrologic cycle (Figure 13.2) and how streams interact with the groundwater table (Figures 13.8 and 13.11).

During Lecture

- Keep the "big picture" in mind. During this lecture you want to learn how streams work. You will learn how water flows in currents, how streams break

up and erode solid rock, and how streams channels and entire drainage systems evolve over time.
- In this lecture it will be particularly helpful to have the preview questions in front of you during the lecture. Try to keep track of which preview each segment of the lecture is addressing.
- It may be helpful to bookmark some of the key figures so you can refer to them and annotate your textbook as they are discussed during lecture. The following are almost certain to be addressed: Figure 14.7 (Channel flow), Figure 14.10 (Formation of natural levees by river floods), and Figure 14.20 (Typical drainage networks). Any figure from the text that is discussed in detail during lecture should be bookmarked and promptly sketched in your lecture notes. Be sure to leave space in your notes to add sketches of textbook figures. If you are not good at sketching or run out of time, cut and paste a paper copy of the figure into your notes.

Note-Taking Tip: Leave Plenty of Room for Visual Material

There is a lot of visual material in this lecture. It will be important to take notes in a format that will allow plenty of room to go back after lecture and make sketches that will help you understand what you wrote in your notes. If you take notes in a loose-leaf notebook leaving room is easy. Just open your notebook, take notes on the right page, leave the left page blank. Then, you can go back after lecture and add simplified sketches on the blank left page. Use the text figures to help you remember what the figures looked like. If you take notes in a spiral notebook you can divide the page into two columns by drawing a vertical line. Take notes in the right column; leave the left column for sketches.

After Lecture

The perfect time to review your notes is right after lecture. The following checklist contains both general review tips and specific suggestions for this chapter.

Check your notes: Have you...

☐ explained how streams flow?
☐ explained how sediment loads are moved in water?
☐ explained how stream valleys and their floodplains evolve?
☐ explained the relationship between stream slope, velocity, discharge, and sediment transport?
☐ explained how drainage systems work?
☐ filled in from memory anything you didn't have time to write down during the lecture?
☐ added visual material to your notes? *Suggestion: Be sure to include simple sketches of the types of drainage networks with an annotation about the factor(s) that influence the formation of each type.*

Intensive Study Session

Set priorities for studying this chapter. We recommend you give highest priority to activities that involve answering questions. We recommend the following strategy for learning this chapter.

- **Text.** First and foremost, review the key figures in Chapter 14. You have to understand these figures to answer the other questions: Figure 14.7 (Channel flow), Figure 14.9 (Channel patterns), Figure 14.10 (The formation of natural levees by river floods), Figure 14.13 (The longitudinal profile), Figure 14.15 (A Change in base level), Figure 14.20 (Typical drainage networks), Figure 14.21 (Antecedent stream), and Figure 14.22 (Superposed stream). This is a long list of key figures, so feel free to go right to the questions and refer to the figures as you answer them. An excellent review strategy is to answer all of the Exercises at the end of the chapter sometime before your next exam. These are short answer and won't take long if you know the material.
- **Practice Exercises and Review Questions.** Next, complete Exercise 1: Stream Velocity. This is an easy way to master one of the most important ideas in the chapter. Then, try the other **Practice Exercises and Review Questions.** Check your answers as you go, but try to answer the question before you look at the answer.
- **Text Exercises.** These exercises are at the end of Chapter 14. Sometime before your next exam answer all of the exercises at the end of the chapter. These are short answers and won't take long if you know the material.
- **Web Site Study Resources**
  http://www.whfreeman.com/understandingearth
  Complete the **Concept Self-Checker** and **Web Review Questions.** Pay particular attention to the explanations for the answers. **Flashcards** will help you learn the terms in this chapter. Do the **Online Review Exercises:** *Identify the Parts of a Marine Delta, Flooding,* and *Create a Flood.*

  Albert Einstein once asked the question: Why do rivers meander? Figure Story 14.9 illustrates steps in the formation of river meanders. Why meanders form is still debated. The **Geology in Practice** exercises explore ideas for the formation of river meanders and their characteristics.

Mark Twain, noting how muddy the Missouri was, proclaimed it "too thick to navigate, but too thin to cultivate."

Exam Prep

Materials in this section are most useful during your preparation for midterm and final examinations.

> **Exam Prep Tip**
>
> Study Chapters 13 and 14 as a package. Review them as one integrated unit. Think of the ways in which groundwater and stream flow are linked.

The following **Chapter Summary** and **Practice Exercises and Review Questions** should simplify your chapter review. Read the **Chapter Summary** to begin your session. It provides a helpful overview that should refresh your memory.

Next, work on the **Practice Exercises and Review Questions.** In order to determine how you stand on mastery of this chapter, complete the exercises and questions just as you would a midterm. After you answer the questions score them. Finally, review any question that you missed. Identify and correct the misconception(s) that resulted in your answering the question incorrectly.

Chapter Summary

How does flowing water in streams erode solid rock and transport and deposit sediments?

- Streams erode, transport, and deposit sediments. Turbulent stream flow allows water to erode and transport sediment by suspension, saltation, rolling, and sliding. The tendency for particles to be carried in suspension is countered by gravity, pulling them to the bottom, and measured by the settling velocity. Deposition of sediments occurs when the velocity of the stream decreases. Refer to Figures 14.2 and 14.3.

How do stream valleys and their channels and floodplains evolve?

- The physical features (drainage pattern, stream channel, floodplain, meander bends in the channel, alluvial fans, and deltas) of a stream system evolve over time.
- The longitudinal profile represents the stream gradient. It is a plot of the elevation of the stream channel bottom at different distances along the stream's course. The longitudinal profile is controlled by local (the river into which the stream flows or a lake) and regional (the ocean) base levels. Refer to Figure 14.14. Streams cannot cut below base level, because base level is the bottom of the hill.

How is the stream's gradient (slope), velocity (speed), discharge (amount of water), and sediment transport linked?

- Whether a stream is dominantly eroding or depositing its load (sediments) is determined by stream velocity. Stream velocity in turn depends on the stream's gradient (slope), discharge (amount of water in the stream), load (sediment in transport), and channel characteristics. A stream's drainage patterns, the stream channel, and floodplain evolve in response to changes in stream velocity, gradient, sediment load, discharge, and characteristics of the landscape over which the stream flows. Alluvial fans form at mountain fronts, in response to an abrupt widening of the stream valley and a change in slope.

How do drainage networks work as collection systems and deltas as distribution systems for water and sediment?

- Drainage networks exhibit different patterns depending on topography, rock type, and geologic structure in the drainage area. Near its mouth, a river tends to branch downstream into distributary channels forming a delta. Deltas are major sites of sediment deposition. Where waves, tides, and shoreline currents are strong, deltas may be modified or even absent. Tectonics controls delta formation by uplift in the drainage basin and subsidence in the delta region.

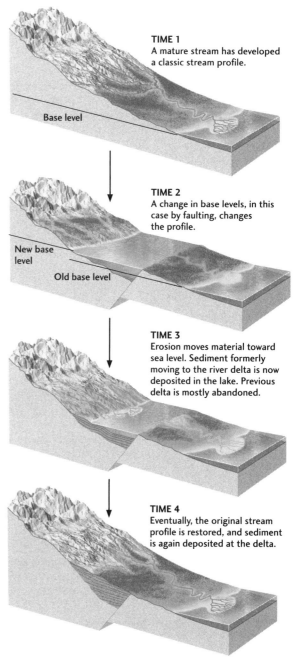

Figure 14.14. The base level of a stream controls the lower end of its longitudinal profile. The profiles illustrated here are for natural regional and local base levels of a river flowing into a lake and from the lake into an ocean. In each river segment, the profile adjusts to the lowest level that the river can reach.

Practice Exercises and Review Questions

Answers and explanations are provided at the end of this Study Guide.

Exercise 1: Stream Velocity

Stream velocity is a dependent variable that governs stream behavior—whether a stream is dominantly eroding and transporting or depositing sediments along a section of the channel.

Various independent variables (factors) influence stream velocity, and therefore, can affect the behavior of a certain stretch of stream channel. The major independent variables affecting velocity are listed in the table below. Complete the table by describing how changes in each factor affects stream velocity.

Variable affecting stream velocity	Relationship of variable to stream velocity	Analogy
Gradient–the slope of the stream channel		Do you tend to walk faster down a steeper slope or a more gradual slope?—faster down a steeper slope.
Discharge–the amount of water in the stream channel		Will you move into a new house slower or faster if you have more people helping you?—faster.
Sediment load		Typically, will you travel faster or slower if you are carrying a heavier load?—slower.
Channel characteristics:		
• Channel roughness	*As channel roughness increases, velocity will decrease.*	*Cross-country hiking without a trail tends to slow one down.*
• Channel shape	*The stream has more contact with the channel surface if the channel shape is very wide or very narrow. More contact with the channel will increase drag and decrease velocity.*	*When you have more contact with the ground surface (for instance, when you crawl on your knees), you move slower than when you have less contact with the ground surface (for instance, when you are walking upright on your feet).*

Exercise 2: Relationship Between Stream Flow and Groundwater

Why do streams in desert regions typically flow intermittently and streams in more temperate regions, such as New England, flow year-round? Drawing well-labeled diagrams illustrating each situation with a brief discussion is an excellent way to answer this question. Refer to Figure 13.2.

Desert (Ephemeral) Stream	Temperate (Perennial) Stream

Exercise 3: How Do Rivers Cut Through Mountain Ranges?

The Kali Gandaki River cuts one of the deepest gorges on Earth right through the Himalayan Mountains. Briefly describe two ways that a river can cut through a mountain range. Refer to Figures 14.21 and 14.22.

A. _____

_____

_____

B. _____

_____

_____

Review Questions

Answers and explanations are provided at the end of this Study Guide.

1. Where do rivers obtain their power to erode and transport sediments?
   A. Heat
   B. Gravity
   C. Electricity
   D. Friction

2. The volume of water that flows past a given point along the stream channel in a given interval of time is the
   A. velocity.
   B. discharge.
   C. capacity.
   D. gradient.

3. If the gradient (slope) of a stream is increased, what happens to the velocity of the water?
   A. Velocity increases.
   B. Velocity decreases.
   C. Velocity remains unaffected.
   D. Velocity may increase or decrease.

4. Stream erosion and deposition are primarily controlled by a river's
   A. width.
   B. velocity.
   C. depth.
   D. channel shape.
   Hint: Refer to Figure 14.3.

5. You are canoeing a river in remote Alaska and your GPS reads an elevation of 2500 feet. After paddling for 5 days, you calculate that you have traveled 200 miles. Your GPS now reads an elevation of 2300 feet. What is the stream's gradient in feet/mile for the stretch you just canoed?
   A. 1 foot/mile
   B. 2 feet/mile
   C. 5 feet/mile
   D. 10 feet/mile

6. A trellis drainage pattern forms on
   A. horizontal lava flows.
   B. folded and tilted sedimentary rock layers of varying resistance.
   C. a dome.
   D. horizontal sedimentary rocks.
   Hint: Refer to Figure 14.20c.

7. Stream competence is measured by
   A. the largest particle size that the stream can transport.
   B. the amount of material in the dissolved load.
   C. the maximum width of the channel along the floodplain.
   D. the total amount of suspended and bed load.

8. Particles tend to settle out of the suspended load in the following order:
   A. clay, sand, pebbles.
   B. pebbles, sand, silt.
   C. sand, pebbles, cobbles.
   D. clay, pebbles, sand.

9. Active erosion in a meander bend takes place
   A. in the center of the stream.
   B. along the outer bank of a bend.
   C. along the inside bank of a bend.
   D. near a stream's headwaters.

10. Entrenched meanders, such as those shown for the San Juan River in Figure 14.8, are evidence of
    A. a decrease in stream gradient.
    B. a decrease in discharge.
    C. a change, such that the river has renewed ability to erode.
    D. a decrease in stream velocity.

11. If a dam is placed across a stream that has been carrying a large volume of sediment, the stream would probably
    A. deposit downstream and erode upstream from the dam.
    B. erode downstream and not change upstream from the dam.
    C. erode downstream and gradually deposit upstream from the dam.
    D. not change downstream but would deposit upstream from the dam.
    Hint: Refer to Figure 14.15.

12. If crushed rock and coal waste are regularly dumped into a stream, this leads to
    A. erosion upstream and deposition downstream from the dump.
    B. erosion both upstream and downstream from the dump.
    C. deposition both upstream and downstream from the dump.
    D. deposition upstream and erosion downstream from the dump.

13. If flood control engineers straighten out a meandering stream channel, the stream will probably
    A. flow more slowly in the straightened stretch.
    B. deposit in the straightened stretch.
    C. downcut in the straightened stretch.
    D. downcut and flow more rapidly in the straightened stretch.

14. If a flood is classified as a 50-year flood, it is meant that
    A. it has been 50 years since the last flood that large occurred.
    B. a flood at least that large has occurred every year within the last 50 years.
    C. the flooded area is safe from a serious flood for at least 50 years.
    D. a flood that large occurs on the average of once every 50 years but also has a chance of occurring during any year.
    Hint: Refer to Figure 14.12.

15. Stream velocity generally increases downstream even though there is a decrease in stream gradient because
    A. the river channel typically gets rougher downstream.
    B. channels typically meander less downstream.
    C. the amount of sediment decreases downstream.
    D. stream discharge typically increases downstream as tributaries contribute their water.

16. If the regional base level of a stream is lowered
    A. the stream will deposit to raise the base level to its former position.
    B. the stream will begin to downcut at its headwaters.
    C. the stream will begin to downcut at its downstream end, and downcutting will progress upstream until the stream channel is graded with the new base level.
    D. a change in base level will have no effect on the stream.

17. Where streams emerge from a narrow mountain canyon onto a flat plain, alluvial fans will form because
    A. the increase in the amount of water from tributary canyons results in deposition.
    B. stream velocity decreases due to a widening of the stream channel and a decrease in gradient.
    C. stream velocity increases due to a decrease in gradient.
    D. all of the above.
    Hint: Refer to Figure 14.16.

18. What stream feature(s) can develop as a result of regional uplift and erosion?
    A. Accelerated downcutting in streams
    B. Stream terraces
    C. Incised meanders
    D. All of the above
    Hint: Refer to Figures 14.8 and 14.17.

19. Which house in the illustration below will ultimately fall into the river channel due to channel bank erosion?
    A. House A   C. House C
    B. House B   D. House D
    Hint: Refer to Figure Story 14.9.

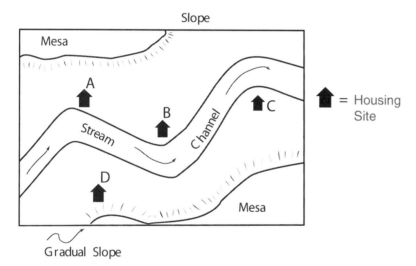

CHAPTER 15

Winds and Deserts

The wind grew stronger, whisked under stones, carried up straws and old leaves, and even little clods, marking its course as it sailed across the fields. The air and sky darkened, and through them the sun shone redly, and there was a raw sting in the air.

—John Steinbeck, The Grapes of Wrath, 1939

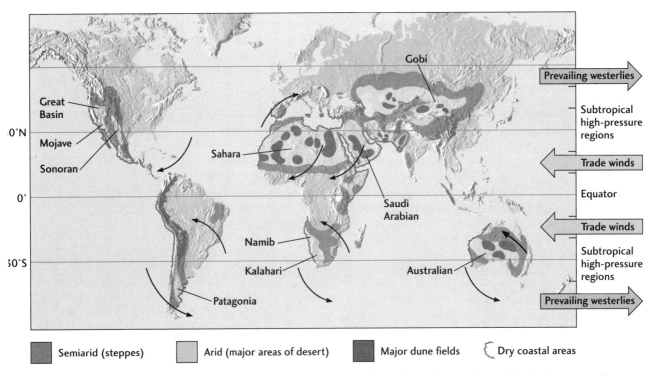

Figure 15.15. Major desert areas of the world (exclusive of polar deserts) in relation to prevailing wind directions and major mountain and plateau areas. Sand dunes make up only a small proportion of the total desert area.

Before Lecture

Before you attend the lecture be sure to spend some time previewing Chapter 15. For an efficient preview, use the following questions. **Chapter Preview** questions constitute the basic framework for understanding this chapter. Preview works best if you do it just before lecture. With the main points in mind you will understand the lecture better. This in turn will result in a better and more complete set of notes.

Chapter Preview

- **Where do winds form and how do they flow?**
  Brief answer: Warm air rises at the equator, causing cloudiness and abundant rain in the tropics. Air flows toward the poles and at about 30 degrees north and 30 degrees south the cooled air sinks, warms up, absorbs moisture, and often produces dominantly clear skies and deserts. Refer to Figures 15.1 and 15.15.
- **What factors contribute to the existence of desert regions on Earth?**
  Brief answer: Global atmospheric circulation patterns, distance from water (oceans), and topographic barriers, such as mountains, cold ocean currents, and polar climates, contribute to the formation of desert conditions on Earth.
- **How do winds erode and transport sand and finer-grained sediments?**
  Brief answer: Turbulent airflow and forward motion combine to lift finer particles into the wind and carry them by suspension, sliding, rolling, and saltation.
- **How do winds deposit sand dunes and dust?**
  Brief answer: A decrease in wind velocity and gravity causes deposition of sediment being transported by wind. The formation and shape of sand dunes depend on the supply of sand and the strength and variability of wind direction.
- **What features are characteristic of desert landscapes?**
  Suggestion: Many diagnostic features of deserts are described in this chapter. One question to ask yourself as you read about these features is, "What features might a geologist look for in the rock record that would serve as evidence for past desert conditions at localities that are no longer deserts?"

How Much Time Do You Have Before Lecture Begins?	How to Use It:
30 minutes or more	With this much time you can dig deep into the chapter. Do as many of the following as your time allows. ✓ Read the **Chapter Preview** questions and brief answers. ✓ Read the suggestions for **During Lecture.** ✓ Study the key figure(s) for this chapter (usually shown at the beginning of the Study Guide chapter). ✓ Study and annotate any additional figures, hints, or suggestions alluded to in the **Chapter Preview.** ✓ If time allows do the **Practice Exercises and review questions.**
15–20 minutes	Do a brief but intense preview: ✓ Read the **Chapter Preview** questions and brief answers. ✓ Read the suggestions for **During Lecture.** ✓ Study the key figure for this chapter (always shown at the beginning the Study Guide chapter).
5–10 minutes	Read the **Chapter Preview** questions and brief answers. Focus on getting the questions clearly in mind. Then listen for answers during lecture. Even 5 minutes of previewing helps!

Vital Information from Other Chapters

Figure 7.16 illustrating a typical desert soil profile and associated text on *Dry Climates: Pedocals* is important information to review in Chapter 7. Also, review the section *Hydrology and Climate* in Chapter 13.

During Lecture

One goal for lecture should be to leave class with a good set of answers to the preview questions.

- To avoid getting lost in details, keep the "big picture" in mind. Chapter 15 tells two related stories. First, the story of wind: Earth's atmospheric circulation pattern and how wind transports sediment and creates sand dunes. Second, the story of deserts: how Earth's circulation patterns produce deserts 15 to 30 degrees away from the equator and how unique features of a desert landscape (e.g., desert pavement, sand dunes, and pediments) evolve.
- Focus on understanding Figures 15.1 (atmospheric circulation) and 15.12 (dune types).

After Lecture

The perfect time to review your notes is right after lecture. The checklist below contains both general review tips and specific suggestions for this chapter.

Check your notes: Have you...

- ☐ added your own sketches of the four types of sand dunes?
- ☐ created a brief "big picture" overview of this lecture (using a sketch or written outline)? *Suggestions for Chapter 15: Sketch a simple figure that integrates the information in Figures 15.1 and 15.15. Your sketch should answer the question: "Why do deserts tend to occur between 15 to 30 degrees on either side of the equator?"*

Intensive Study Session

Set priorities for studying this chapter. We recommend you give highest priority to activities that involve answering questions. We recommend the following strategy for learning this chapter.

- **First, preview the key figures in chapter.** Focus your attention on understanding Earth's atmospheric circulation pattern (Figure 15.1 and how this is related to the formation of deserts, as shown in Figure 15.15). Then, move on to understanding other features of deserts. Pay particular attention to the types of sand dunes (Figure 15.12), and the formation of mountain pediments (Figure 15.20).
- **Next, complete the Practice Exercises and Review Questions.** Exercise 1: Sand Dune Types will help you remember the different types of sand dunes. Then, try answering each of the review questions to check your understanding of the lecture. Check your answers as you go, but do try to answer the question before you look at the answer.
- **Web Site Study Resources**
  http://www.whfreeman.com/understandingearth
  Complete the **Web Review Questions.** Pay particular attention to the explanations for the answers. Do the **Online Review Exercises:** *Identify the World's Major Deserts* and *Understanding Global Air Circulation Patterns*. Check out images of deserts and desert features in the **Photo Gallery.** Explore the link between dust storms, rainforests, and coral reefs by doing the **Geology in Practice** exercises.

Exam Prep

Materials in this section are most useful during your preparation for exams. The following **Chapter Summary** and **Practice Exercises and Review Questions** should simplify your chapter review. Read the **Chapter Summary** to begin your session. It provides a helpful overview that should refresh your memory.

Next, work on the **Practice Exercises and Review Questions.** In order to determine how you stand on mastery of this chapter, complete the exercises and questions just as you would a midterm. After you answer the questions score them. Finally, review any question that you missed. Identify and correct the misconception(s) that resulted in your answering the question incorrectly.

Chapter Summary

Where do winds form and how do they flow?

- Prevailing winds on Earth are largely controlled by zones of rising and sinking air and the Coriolis effect. Hot (less dense) air rises and carries moisture up to where it condenses and falls as precipitation. As air radiates heat to space in the upper atmosphere, it cools, becomes denser, and sinks to the surface and flows back toward the equator. Sinking air is typically very dry because it comes from the cold and dry upper atmosphere. Due to the rotation of Earth, the Coriolis effect deflects air flow in both hemispheres.

What factors contribute to the existence of desert regions on Earth?

- Deserts are regions where evaporation exceeds precipitation.
- Desert regions on Earth are the result of: 1) global air circulation patterns, which generate a relatively stationary zone of descending, warm, dry air at about 15° to 30° north and south of the equator; 2) distance from large bodies of water (e.g., oceans); 3) the rain shadow generated by high mountains block the flow of moisture-rich air; 4) cold ocean currents which reduce air temperatures and reduce the transport of moisture by the air; and 5) polar climates where the air is so cold that it holds very little moisture at all.

How do winds erode and transport sand and finer-grained sediments?

- Wind is a major erosional and depositional agent, moving enormous quantities of sand, silt, and dust. Turbulent air flow within wind erodes and transports particles by suspension, sliding, rolling, and saltation. As velocity decreases, sediment is pulled out of air by gravity and deposited as a blanket or dunes of sand and dust.

How do winds deposit sand dunes and dust?

- Only about 20% of the area of desert regions are covered by sand. The distinctive types of sand dunes are governed by the amount of sand available, the strength of the wind, and the variability in wind direction. Loess (wind blown dust) is another important wind deposit.

What features are characteristic of desert landscapes?

- Geologic and topographic features associated with desert regions include: sand dunes, loess deposits, evaporite (salt) deposits, desert pavement, ventifacts (sandblasted rocks), alluvial fans and alluvial sands and gravels, pediments, mesas, distinctive soils (pedocals, rich in calcium carbonate) and rusty orange-brown colors to weathered rock surfaces. Some of these features are preserved in the rock record and provide geologists clues to ancient desert regions that no longer exist.

Practice Exercises and Review Questions

Answers and explanations are provided at the end of this Study Guide.

Exercise 1: Sand dune types

Complete the table below by filling in the blank spaces. Figure 15.12 and the section in your text on *Dune Types* will be helpful. Also refer to pictures of dunes in the **Photo Gallery** on the web site.

Dune type	Characteristics	Sand supply	Wind direction/strength
Barchan			
Transverse			
Blowout		Limited to moderate	Unidirectional/gusty
Linear (See Figure 15.8)			

Review Questions

Answers and explanations are provided at the end of this Study Guide.

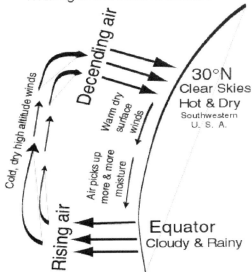

A Hadley Cell — atmospheric circulation pattern generated by the differential heating of the Earth's surface

1. Wind belts on Earth are largely controlled by
   A. regions of sinking air.
   B. regions of rising air.
   C. the Coriolis effect.
   D. all of the above.

2. Deserts may be caused by all of the following except
   A. rising air.
   B. proximity to cold ocean currents.
   C. great distance from the ocean.
   D. descending air.

3. As a dune advances
   A. sand erodes from the windward slope.
   B. sand is deposited on the leeward slope.
   C. particles move over the crest by saltation.
   D. All of the above.

4. As a dune grows in height wind streamlines over the dune become
   A. less compressed and their velocity decreases.
   B. less compressed and their velocity increases.
   C. more compressed and their velocity decreases.
   D. more compressed and their velocity increases.

5. A dried lake bed with abundant salt deposits and flat enough for the Space Shuttle to land on is called
   A. pediment.
   B. desert pavement.
   C. a playa.
   D. an oxbow lake.

6. Loess is
   A. fine dust transported by wind in suspension and deposited on land.
   B. dust transported by wind, deposited, and then later eroded and redeposited by water.
   C. fine sand transported and deposited by wind.
   D. fine-grained salt particles eroded by wind off a playa surface.

7. Desert pavement may be the product of
   A. deposition of gravel by flash floods.
   B. concentration of coarser particles as wind removes finer material.
   C. deposition of mud by flash floods.
   D. desert varnish.
   Hint: Refer to Figure 15.7.

8. Rock material carried in suspension by wind is mostly the size of
   A. sand.   C. gravel.
   B. silt and clay.   D. pebbles.

9. The Coriolis effect deflects winds in the northern hemisphere to the
   A. left (westward).   C. north.
   B. right (eastward). D. south.
   Hint: Refer to Figure 15.1 and the section titled *Wind Belts* in the text.

10. You are lost in the Goblin Desert. The nearest town is due south. Winds blow from the south to the north but there is no wind blowing today. Furthermore, the sun is totally obscured by clouds. Without a compass, which direction are you going to hike, given the orientation of the barchan dune shown below?

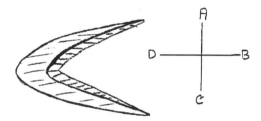

   A. Direction A
   B. Direction B
   C. Direction C
   D. Direction D
   Hint: Refer to Figure 15.12.

11. While hiking through a dune-filled coastal plain on a windless morning, you become surrounded by a dense fog and realize you are lost. You know you are near a shoreline and that the beach will lead you back to camp, but you don't know which direction it is in. You recognize the crescent-shaped dunes that wrap moderate depressions as blow-out dunes! According to your compass, the tapered arms of the dunes point south. Then, remembering that in this region strong, gusty winds come onto the coastal plain off the ocean, you immediately remember which direction to head to get to the beach:
    A. North   C. East
    B. South   D. West
    Hint: Refer to Figure 15.12

12. Why are evaporites significant geological deposits?
    A. They are a good paleoenvironmental indicator of ancient desert conditions.
    B. They are a major source for chemicals, such as borax.
    C. They are a source of salt for the dinner table.
    D. All of the above.

13. Chemical weathering in desert regions is slow due to the lack of water. Therefore, desert soils form very slowly and are called _____ , which are enriched in _____ .

    Hint: Refer to Figure 7.16.

14. Petroglyphs, early Native American artwork, is scratched through _____ .

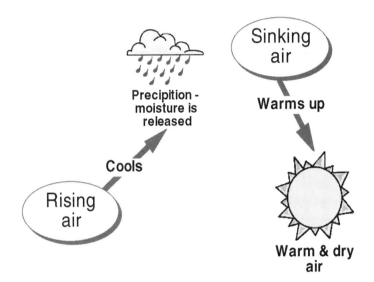

CHAPTER 16

Glaciers: The Work of Ice

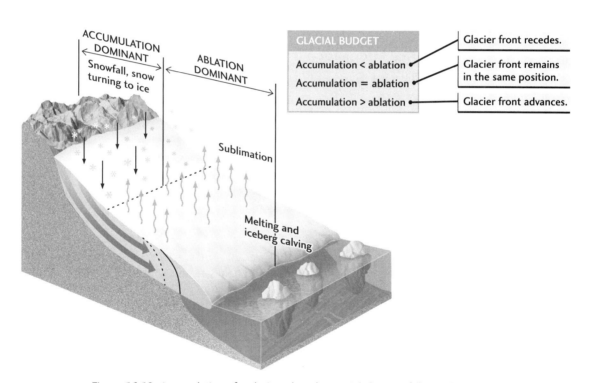

Figure 16.10. Accumulation of a glacier takes place mainly by snowfall over the colder upper regions. Ablation takes place mainly in the warmer lower regions by sublimation, melting, or iceberg calving. The difference between accumulation and ablation is the glacial budget.

Before Lecture

Before you attend lecture be sure to spend some time previewing the chapter. For an efficient preview, use the questions below.

Chapter Preview

- **How do glaciers form and how do they move?**
  Brief answer: Glaciers form where snow accumulation exceeds melting. Refer to Figure 16.10. Glaciers move by a combination of plastic flow and slip at the base of the ice. See Figure 16.12.
- **How do glaciers erode bedrock, transport and deposit sediments, and shape the landscape?**
  Brief answer: Glaciers erode by scraping, plucking, and grinding rock. Refer to Figure 16.18. The rock debris deposited by a glacier is called till. Till may contain rock flour to giant boulders. Glaciers sculpture a distinctive landscape with U-shaped and hanging valleys, arêtes, cirques, moraines, drumlins, kames, and other features. Refer to Figures 16.17, 16.19, 16.20, and 16.21.
- **What are the ice ages and what causes them?**
  Brief answer: During the Pleistocene many cycles of advance and retreat of continental icesheets occurred. The ice sheets of the last major advance were gone by about 10,000 years ago, the beginning of the Holocene Epoch. A popular theory for the cause of the Pleistocene ice ages relies on how variations in the Earth's orbit affect the intensity of solar energy reaching Earth. Refer to Figure 16.27.
- **Why do ice shelves float?**
  Brief answer: Ice floats on the ocean in exactly the same way that ice cubes float in a glass of water. Ice is less dense than water. See Figure Story 16.16.

Vital Information from Other chapters

Review Figure 13.1 (The distribution of water on Earth).

During Lecture

- You know by now that it helps with note taking if you can keep the "big picture" in mind. This chapter tells the story of ice: how glaciers form and move and create a landscape of unique features. This story has a deep history: the many ice ages of the Pleistocene Epoch. Finally, the chapter suggests a future when our descendants will almost certainly have to face the return of ice and massive glaciation.
- Note that understanding glacial landscapes is one of the main goals for this lecture. Take notes that will help you understand key landscape features, such as the following:

   Glacial features formed by the erosive power of glacial ice (refer to Figures 16.17, 16.18, and 16.19):
   - striations–scratches and grooves–carved in bedrock over which the glacier flowed.
   - cirque
   - U-shaped valley
   - hanging valley
   - fjord
   - arête

   Glacial features formed by deposition of rock material by glacial ice (refer to Figures 16.20 and 16.21):
   - glacial moraines—the different types are described in Table 16.1.
   - drumlins
   - esker
   - kame

- glacial erratic
- kettle
- varves

For each feature above try to sketch essentials and annotate what makes this feature unique. Write figure numbers of slides taken from the text in the margin so you can review them later.

After Lecture

Check your notes: Have you...

- ☐ captured the glacial processes? Your notes should say clearly: 1) how glaciers form and move; and 2) how they erode material.
- ☐ described the essential features of the glacial landscape? Refer to **During Lecture** (above) for a list of landscape features you should have in your notes on the **During Lecture** section. Be sure to go back to the text for features you missed.
- ☐ added visual material? Doing simple sketches of the glacial landscape features is a great way to learn them. It will be helpful to have all your features on a single page in your notes for easy review.

Intensive Study Session

Study Tip

The pictures in Chapter 16 are designed as a virtual field trip. Use the pictures to master glacial landscape features. Study each picture until you understand what it is and how it differs from other features. (For a list of the landscape features you need to study, refer to the **During Lecture** section).

Set priorities for studying this chapter. We recommend you give highest priority to activities that involve answering questions. We recommend the following strategy for learning this chapter.

- **First, preview the key figures in the chapter.** You will get the greatest return on your study time by working on these figures because they will help you remember the most important ideas in the chapter. Figures 16.10 and 16.12 will help you understand how glaciers form and move. Figure 16.18 explains how glaciers erode bedrock. Figures 16.17, 16.19, 16.20, and 16.21 and Table 16.1 provide a virtual field trip of the distinctive features of the glacial landscape. Figure Story 16.16 explains why ice shelves float.
- **Then, try the Practice Exercises and Review Questions.** Start with Exercise 1, which will help you learn the distinctive features of glacial landscape. Then go to the multiple choice **Review Questions** to check your understanding of the lecture. Try to answer each question before you look at the answer.
- **Sometime before your exam, answer all 11 of the exercises at the end of the chapter in the text.** These are short answer and won't take long if you know the material. They make an excellent review of Chapter 16.
- **Web site Online Review Exercises and Study Tools**
  http://www.whfreeman.com/understandingearth
  Complete the **Concept Self-Checker** and **Web Review Questions.** Pay particular attention to the explanations for the answers. Explore the **Photo Gallery** for images of landforms produced by glacial ice. The **Flashcards** will

help you review the key terms in the chapter. Do the **Geology in Practice** exercises for more practice with the concepts covered in this chapter.

Exam Prep

Materials in this section are most useful during your preparation for midterm and final examinations. The following **Chapter Summary** and **Practice Exercises and Review Questions** should simplify your chapter review. Read the **Chapter Summary** to begin your session. It provides a helpful overview that should refresh your memory.

Next, work on the **Practice Exercises and Review Questions.** In order to determine how you stand on mastery of this chapter, complete the exercises and questions just as you would a midterm. After you answer the questions score them. Finally, review any question that you missed. Identify and correct the misconception(s) that resulted in your answering the question incorrectly.

Chapter Summary

How do glaciers form and how do they move?

- Glaciers form in cold and snow climates where snow accumulation exceeds the ablation of ice due to melting, sublimation, wind erosion, and iceberg calving. Glacial ice moves by plastic flow and slip along the base, which may be lubricated by melt water. The rate of ice flow varies typically from meters per year to meters per week.
- Glaciers are described as advancing or retreating depending on the balance between snow accumulation and ablation. When ablation exceeds accumulation, the shrinking glacier "retreats" as the toe or terminus moves up slope. When accumulation exceeds ablation, the expanding glacier "advances" as its toe or terminus moves down slope.

How do glaciers erode bedrock, transport and deposit sediments, and shape the landscape?

- Glaciers are powerful agents of erosion and deposition. Glaciers erode by scraping, plucking, and grinding rock.
- Landscapes sculptured by ice have distinctive features that have provided geologists with evidence for reconstructing the position of ice sheets during the ice ages and deciphering the existence of ice ages throughout Earth's history. U-shaped and hanging valleys, moraines, arêtes, cirques, drumlins, kames, eskers, striated rock, and other features characterize a glacial landscape.
- Ice-laid deposits of rock material are called till, consisting of heterogeneous mixture of rock, sand, and clay. Accumulations of till are called moraines; each type of moraine is named for its position relative to the glacier that formed it. Ancient tills, called tillites, provide evidence for ancient glaciations numerous times during Earth's history.
- Water-laid deposits from glaciers are called outwash and consist of sand, gravel, and fine rock flour.

What are the ice ages and what causes them?

- The ice sheets of the last major advance were gone by about 10,000 years ago, the beginning of the Holocene Epoch. Studies of the geologic ages of glacial deposits on land and sediments of the seafloor show that the Pleistocene glacial epoch consisted of multiple advances (glacial intervals) and retreats (interglacial intervals) of the continental ice sheets. Each advance corresponded to a global lowering of sea level that exposed large areas of continental shelf; during the interglacial intervals, sea level rose and submerged the shelves.

- Although the causes of the ice ages remain uncertain, the general cooling of the Earth leading to glaciation appears to have been the result of plate tectonics that gradually moved continents to positions where they obstructed the general transport of heat from the equator to polar regions.
- A favored explanation for the alternation of glacial and interglacial intervals is the effect of astronomical cycles, by which very small periodic changes in Earth's orbit and axis of rotation alter the amount of sunlight received at the Earth's surface. There is also evidence that decreased levels of carbon dioxide in the atmosphere diminished the greenhouse effect, which would contribute to global cooling.

Why do ice shelves float?

- Figure Story 16.16 illustrates how the dynamic balance, called isostasy, between the forces of gravity and buoyancy keep ice shelves and bergs afloat in water. The mass of the floating ice is equal to the mass of the water the iceberg displaces. When the ice melts, it simply replaces the water it displaces and, therefore, there is no change in sea level.

Practice Exercises and Review Questions

Answers and explanations are provided at the end of this Study Guide.

Exercise 1: The Glacial Sculptured Landscape

You are hired as a seasonal ranger at a national park with a spectacular glacially sculptured landscape. However, there is not one glacier in the park today. Your job for the summer is to lead natural history hikes during which you interpret the evidence for the past glaciation that helped formed the park landscape. Briefly describe five features that you can point out along the nature trail that speak to the past glacial episode in this region. Remember a picture/sketch is worth a "thousand words," so a good way to provide a brief description is with a well-labeled diagram.

Features formed by the erosive power of glacial ice:

1. _____

2. _____

3. _____

Features formed by deposition of rock material by glacial ice:

4. _____

5. _____

Exercise 2: Your Personal Budget as a Metaphor for a Glacial Budget

Discuss how the changing balance of cash in your checking account is a good metaphor for an advancing vs retreating glacier.

Exercise 3: Glacial Advances and Retreats

A glacier advances, halts, and retreats. Will the glacier continue to deposit material at its snout while it is halted, and even while it is retreating? Discuss.

Review Questions

Answers and explanations are provided at the end of this Study Guide.

1. Glacial ice is most like
    A. an igneous rock.
    B. a sedimentary rock.
    C. a metamorphic rock.
    D. a sediment.

2. The force which moves glaciers is
    A. recrystallization.
    B. melting.
    C. lubrication.
    D. gravity.

3. As snow is transformed into glacial ice a transitional phase of densely packed granular snow is called
    A. pack ice.
    B. firn.
    C. alpine ice.
    D. crystalline ice.

    Hint: Refer to Figure 16.8.

4. Glaciers retreat when the
    A. accumulation of snow is less than the ablation off the glacier.
    B. accumulation of snow exceeds the ablation off the glacier.
    C. accumulation of snow is equal to the ablation.
    D. boundary between the zone of accumulation and ablation moves to lower elevations.

5. Moraines are
    A. erosional glacial features carved in bedrock over which the ice flowed.
    B. made from glacial outwash sediments.
    C. deposits of loess carried by wind from recently glaciated regions.
    D. deposits of glacial till.

6. A drumlin is a
   A. block of bedrock not quarried away by the bottom of a glacier.
   B. sinuous ridge of water-deposited glacial debris.
   C. small depression formed from the melting of a block of ice buried beneath till.
   D. streamlined hill constructed of glacial till.
   Hint: Refer to Figure 16.21.

7. Ragged, knife-edged ridges are commonly found in glaciated mountains. Such a ridge line is called
   A. a horn.
   B. a cirque.
   C. a col.
   D. an arête.
   Hint: Refer to Figure 16.19.

8. When continental glaciers advance over the land surface
   A. sea level lowers.
   B. sea level rises.
   C. plate tectonic processes are especially active.
   D. Europe is significantly warmer.

9. If all the glacial ice on Earth were to melt, the sea level would be raised by about
   A. 1 meter.
   B. 10 meters.
   C. 65 meters.
   D. 155 meters.
   Hint: Refer to Box 16.2.

10. Milankovitch calculated that the eccentricity of the Earth's orbit varies over a cycle of _____ years.
    A. 10,000
    B. 23,000
    C. 41,000
    D. 100,000
    Hint: Refer to Figure 16.27.

11. The relatively mild climate of Europe, despite its northern latitude, is due to
    A. the release of urban and industrial waste heat.
    B. greenhouse gases over England's industrialized areas.
    C. heat release by ocean currents flowing northeast from equatorial regions.
    D. surface winds that have carried heat from the equator.
    Hint: Refer to Figure 16.28.

12. Orbital factors that impact the Earth's heat budget primarily affect the
    A. reflectivity of the Earth's upper atmosphere.
    B. amount of solar energy reaching our planet.
    C. composition of the atmospheric gases.
    D. distribution of heat over the globe.

13. Which of the following features of a glacial landscape is the product of depositional process, rather than erosion?
    A. Fjord
    B. Kames
    C. Hanging valley
    D. Cirque

14. Which of the following is NOT characteristic of glacial till?
    A. Sand grains of virtually all quartz
    B. Lack of clear stratification
    C. Very poor sorting
    D. Boulder, cobble, pebble, sand, and clay-sized rock particles

15. Ablation refers to the
    A. melting, calving, sublimation, and erosion of glacial ice.
    B. erosion and deposition of glacial sediments.
    C. changes in sea level associated with glacial and interglacial periods.
    D. glacier advance and retreat.

16. The glacial budget is balanced when the rate of accumulation is equal to the rate of
    A. ablation.
    B. deposition.
    C. advance.
    D. retreat.

17. The last major advance of continental ice over North America reached
    A. the Gulf of Mexico.
    B. California.
    C. south of the Great Lakes.
    D. Florida.
    Hint: Refer to Figure 16.24.

18. Which of the following may have influenced Earth's climate during ice ages?
    A. Variations in Earth's orbital characteristics
    B. Changes in the composition of the atmosphere
    C. Plate tectonic movements of the continents
    D. All of the above

19. As a result of the collapse and melting of ice shelves and bergs, such as the Larson Ice Shelf in 2002, sea level
    A. rises.
    B. falls.
    C. fluctuates.
    D. remains unchanged.

CHAPTER 17

Earth Beneath the Oceans

... over all the face of Earth
Main ocean flowed, not idle; but, with warm
Prolifick humour softening all her globe ...

—John Milton, Paradise Lost, Book VII, l 278–280

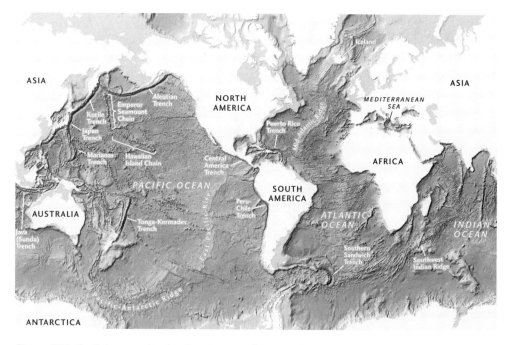

Figure 17.1. Earth topography showing the major features of the ocean floor.

Before Lecture

Chapter Preview

- **How does the geology of the oceans differ from that of the continents?**
  Brief answer: Ocean basins are created at ocean ridges (diverging margins) by volcanism and are destroyed in a brief period (of several hundred million years) by subduction at converging margins.
- **How is deep seafloor formed?**
  Brief answer: The deep seafloor is constructed by basaltic volcanism at the ocean ridges and at ocean hot spots and by deposition of fine-grained clastic and biochemically precipitated sediments. Prominent features of the deep seafloor landscape include seamounts, guyots, and abysmal plains and hills. Refer to Figures 17.4 and 17.6. Oceanic trenches (Figure 17.6) are formed where ocean plates converge.
- **What are the characteristics of a mid-ocean ridge?**
  Brief answer: A rift valley runs along the crest, basaltic volcanism and earthquake activity are common along the ridge crest, and smokers (hydrothermal springs) percolate through cracks on the flanks of the ridges. Refer to Figures 17.5 and 17.7.
- **What are the major components of the continental margins and adjacent ocean floor?**
  Brief answer: Continental margins are flooded portions of the continent. The continental slope marks the edge of the continent and a transition to deeper water and the ocean floor. Turbidity currents transport fine sediments off the continental shelf and onto the adjacent abyssal ocean floor.
- **What kinds of sedimentation occur in and near the oceans?**
  Terrigenous (note similarity the word terrain) *sediments* are muds and sands eroded from the continent and deposited by turbidity currents along the continental shelf. *Biochemical sediments* result from deposition of calcium carbonate from shells and coral reefs. *Open-ocean* (pelagic) *sediments* result from clays and oozes of calcium carbonate and silica shells of microorganisms. *Evaporite sediments* result from intense evaporation in shallow tropical seas. Sediments derived from volcanic ash and lava flows are deposited near subduction zones.
- **What processes shape the shoreline?**
  Brief answer: Waves and tides interact with tectonics to shape the shoreline. Winds blowing over the sea create waves. Waves approaching shallow water along the shoreline are transformed into breakers and refracted into longshore currents and longshore drift, which transports sand along the beach. Tides deposit sediment on longshore flats. Tectonic uplift creates cliffs and headlands, which are smashed by wave action, leaving behind cliff remnants called seastacks. Tectonic subsidence creates areas of long, wide beaches and low-lying coastal plains and sandbars, which may evolve into barrier islands.

Vital Information from Other Chapters

Chapters 2, 5, and 8 contain key information on ocean basins and rock materials associated with marine environments. In Chapter 2, review Figures 2.3, 2.6, 2.9, and 2.14, and *Earth Issue* box 2.1. In Chapter 5 on Igenous Rocks, review pages 103–107. In Chapter 8, review Figure Stories 8.4 and 8.16, Tables 8.2 and 8.4, and *Earth Policy* box 8.1. A quick reread of Chapter 8 would help immensely.

During Lecture

One goal for lecture should be to leave class with a good set of answers to the preview questions. To avoid getting lost in details, keep the "big picture" in mind. Chapter 17 tells the story of the ocean depths. You will learn about various landform features of the deep ocean, continental margins, and the shoreline, and the geological processes that create them.

> **Note-Taking Tip**
>
> There is a lot of new terminology in this chapter. Because the unit is so terminology rich, your lecturer may use terms you are not familiar with. Mark, circle, or underline these in your notes so you can check them out later. Put the abbreviation *def. (define)* in the margin to remind yourself to do this.

> *. . . though we know the sea to be an everlasting terra incognita, so that Columbus sailed over numberless unknown worlds. . . .*
>
> —HERMANN MELVILLE, *MOBY-DICK* (1851)

After Lecture

The perfect time to review your notes is right after lecture. The checklist below contains both general review tips and specific suggestions for this chapter.

> **Check your notes: Have you...**
>
> ☐ added visual material? *Suggestions: Draw a profile sketch similar to Figures 17.4 and 17.6 that shows the major features of the deep ocean floor.*
> ☐ reworked notes into a form that is efficient for your learning style? Visual learners need to see the figures to study and may want to copy key figures and insert them into their notes near the text that explains them. Kinesthetic learners may find it more beneficial to sketch the figures because the act of drawing helps them remember better.

Intensive Study Session

Set priorities for studying this chapter. We recommend you give highest priority to activities that involve answering questions. We recommend the following strategy for learning this chapter.

- **Text.** First, preview the key figures in Chapter 17. The following sequence of figures tells the story of the ocean: its seafloor topography, continental margins, waves, tides, and beaches. You will need a general understanding of these figures in order to complete the **Practice Exercises and Review Questions.** Figures 17.4, 17.5, and 17.6 will help you learn about seafloor topography and landforms. Figures 17.8 and 17.9 explain the geology of continental margins. Figure Story 17.13 explains how wave action (breaking and refraction) works along the shoreline of an ocean. Figure 17.14 explains how the moon and sun produce the tides. Figures 17.17 and 17.18 explain the structure and sand budget of an ocean beach. Sometime before your exam, answer all 10 of the exercises at the end of the chapter. These are short-answer questions and won't take long if you know the material. Note that there is an interesting animation for Exercise 2 that can be found on the **Media Link Questions** section of the web site.
- **Practice Exercises and Review Questions.** Next, complete Exercise 1: *Profile from the Ocean to the Seashore* and Exercise 2: *Passive vs. Active Continental Margins*. You will get the greatest return on your study time by working on these exercises because they will help you remember some of the

most important ideas in the chapter. Then, try the **Review Questions**. Try answering each of the review questions to check your understanding of the lecture. Check your answers as you go, but do try to answer the question before you look at the answer.

- **Web Site Study Resources**
  http://www.whfreeman.com/understandingearth
  Complete the **Concept Self-Checker** and **Web Review Questions.** Pay particular attention to the explanations for the answers. Do the **Online Review Exercises:** *Identify the Parts of a Beach* and *Understanding Waves and Currents*. The **Geology in Practice** exercise on *Why Are the Oceans Salty?* provides a good review of the link between weathering, erosion, and the composition of sea water.

Study Tip

Now is the time to look up those terms you didn't understand. Skim the margin of your notes. (You did mark them (*def*), didn't you?) Your text provides two helpful aids for dealing with terms you don't know. *Key Terms and Concepts* (at the end of each text chapter) lists all new terms and handily provides the page number of a term so you can look it up. Alternatively, use the *Glossary* at the end of the text.

Exam Prep

Materials in this section are most useful during your preparation for quizzes and exams. The following **Chapter Summary** and **Practice Exercises and Review Questions** should simplify your chapter review. Read the **Chapter Summary** to begin your session. It provides a helpful overview that should refresh your memory.

Next, work on the **Practice Exercises and Review Questions.** In order to determine how you stand on mastery of this chapter, complete the exercises and questions just as you would a midterm. After you answer the questions score them. Finally, review any question that you missed. Identify and correct the misconception(s) that resulted in your answering the question incorrectly.

Chapter Summary

How does the geology of the oceans differ from that of the continents?

- Volcanism and sedimentation shape the ocean floor. Folding, faulting, weathering, and erosion play an important part in shaping the continents.

How is the deep seafloor formed?

- Volcanism along ocean spreading centers (divergent plate boundaries) creates new seafloor, which eventually is recycled back into the mantle by subduction at trenches (convergent plate boundaries).
- Basaltic volcanism, induced by pressure-release melting in the upper mantle beneath spreading centers, generates a thin, iron-rich ocean crust, relative to continental crust. Review pages 102–107 in Chapter 5.

What are the characteristics of a mid-ocean ridge?

- A rift valley marks the crest of ocean ridges.

- Hydrothermal springs form on the rift valley floor as seawater percolates through the new, hot oceanic crust.
- Transform faults offset ocean ridges at many places to accommodate different spreading rates along offset segments of the ridge. Refer to Figure 2.5.

What are the major components of the continental margins and adjacent ocean floor?

- Continental margins are flooded portions of the continent. The continental slope and rise mark the edge of the continent and a transition to deeper water and the ocean floor.
- Passive continental margins form where rifting and seafloor spreading carry continental margins away from active plate boundaries. Active continental margins form where oceanic lithosphere is subducted beneath a continent or a transform fault coincides with the continental margin.
- Continental shelves are broad and relatively flat at passive continental margins and are narrow and uneven at active margins.
- Turbidity currents transport fine sediments off the continental shelf and onto the adjacent abyssal ocean floor. Turbidity currents can both erode and transport sediments. Submarine canyons and fans are formed by turbidity currents.
- Coral reefs and atolls are constructed by coral and other marine organisms. The reef construct plays an important role in modulating wave energy and creating a favorable environment for shallow marine life. Refer to Figure Story 8.16 and *Earth Policy* box 8.1.

What kinds of sedimentation occur in and near the oceans?

- In the deep sea fine-grained pelagic terrigenous and biochemically precipitated sediments settle to the seafloor. Foraminiferal oozes, composed of tiny foraminiferal shells, are the most abundant biochemical component of pelagic sediments. Foraminiferal and other carbonate oozes are abundant at depths less than about 4 km. Below a certain depth, called the carbonate compensation depth, carbonate sediments dissolve in deep seawater. Deep ocean water is colder, contains more carbon dioxide, and is under higher pressure. All these factors increase the solubility of carbonate sediments. Silica ooze is produced by sedimentation of the silica shells of diatoms and radiolaria.

What processes shape the shoreline?

- Waves and tides shape the shoreline. Waves are created by the wind blowing over the surface of the water. Ocean tides on Earth are a result of centrifugal force and gravitational forces acting between the Earth, Moon, and Sun.
- The beach is a result of the dynamic balance between waves and longshore currents that erode and transport rock material along the coast within the surf zone and the supply of sand from rivers that deliver sand to the surf zone. Refer to Figure 17.18.
- Longshore currents result from the zigzag movement of water on and off the beach. Waves typically splash onto shore at an angle in part due to wave refraction. The backwash—off the beach—runs down the beach slope at a small but opposite angle to the swash. The net result of this swash and backwash of water on and off the beach slope is a longshore current that transports sand parallel to the beach within the surf zone.

Practice Exercises and Study Questions

Answers and explanations are provided at the end of this Study Guide.

Exercise 1: Profile from the Atlantic Shoreline to the Ocean Floor

A. Fill in the blanks to correctly label this profile. **Hint:** Refer to Figure 17.8a.

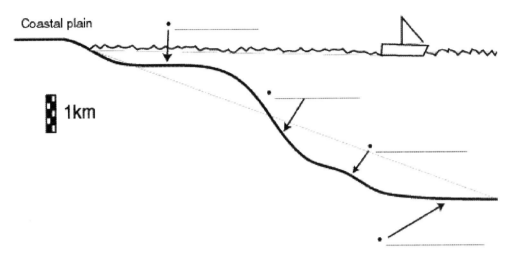

B. Does the profile above illustrate an active or passive continental margin? Explain.

_____

_____

_____

Exercise 2: Passive vs. Active Continental Margins

Characterize each coastal locality described below as either a *passive* or an *active* continental margin. Figures 17.1, 17.4, 17.6, and 17.8 are useful references.

A. A coastline far from an active plate boundary _____

B. East coast of North America _____

C. West coast of South America _____

D. A coastline with a very broad, featureless continental shelf _____

E. California coast along the San Andreas fault _____

F. Continental margin with no volcanic activity for millions of years _____

Review Questions

Answers and explanations are provided at the end of this Study Guide.

1. Ocean waves are mostly generated by
   A. tides.
   B. ships.
   C. the wind.
   D. earthquakes.

2. Geologically, the edge of a continent is considered to be
   A. on the ocean side of the seafloor trenches.
   B. the continental shelf.
   C. the continental rise and slope.
   D. the shoreline.

3. Which of the following processes is most important for building the ocean floor?
   A. Volcanism
   B. Metamorphism
   C. Precipitation of carbonate rocks
   D. Deposition of sediment derived from the land

4. The bending of waves as they approach shore is called
   A. wave reflection.
   B. wave erosion.
   C. longshore drift.
   D. wave refraction.

5. An erosional coast is characterized by
   A. sea cliffs, sea stacks, and wave-cut terraces.
   B. barrier islands.
   C. coral reefs.
   D. estuaries.

6. Ocean floor rock is made up of
   A. basalt and pelagic sediments.
   B. granite and gneiss.
   C. obsidian and sand.
   D. rhyolite and carbonate sediments.

7. The rock that makes up a seamount is
   A. basalt.
   B. limestone.
   C. granite.
   D. marine sedimentary rocks.

8. Sea stacks are
   A. piles of sedimentary rocks near the shore.
   B. the erosional remnants of sea cliffs.
   C. formed where a river drains onto the coastline.
   D. formed by rapidly growing corals on a reef.
   Hint: Refer to Figure 17.12.

9. The deep abyssal ocean plain lies at a water depth of
   A. between 100 to 500 meters.
   B. 600 to 1000 meters.
   C. 2000 to 3000 meters.
   D. 4000 to 6000 meters.
   Hint: Refer to the section titled *Profiles Across Two Oceans* in your text.

10. Recent precise satellite measurements of the sea surface shows that
    A. sea level is rising a few millimeters per year.
    B. sea level is not changing.
    C. sea level is dropping a few millimeters per year.
    D. seal level rises and falls with the seasons.

11. What is sediment-charged water which flows rapidly down the continental slope called?
    A. Turbidity current
    B. Longshore current
    C. Tidal current
    D. Tsunami

12. The highest tides occur when
    A. the Sun, Moon, and Earth are all aligned.
    B. the Sun and Moon are at right angles to the Earth.
    C. the Earth is closest to the Sun.
    D. the Moon is in either its first or last quarters.
    Hint: Refer to Figure 17.14

13. You live on a beach front. Your up-current neighbors are planning to build a groin to halt erosion and increase the width of their beach. What effect will this have on your beach?
    A. The groin will likely cause your beach to grow.
    B. It is likely that your beach will remain unchanged.
    C. The groin will likely cause your beach to erode.
    D. None of the above.
    Hint: Refer to *Earth Policy* box 17.1.

14. Headlands and points experience greater erosion than bays and inlets because of
    A. longshore drift.
    B. wave refraction.
    C. wave reflection.
    D. the more resistant rock in the headland.
    Hint: Refer to Figure Story 17.13.

15. If a river delivers sand to a shoreline faster than currents can transport the sediment, the result is
    A. erosion of the sand to form a rocky coastline.
    B. formation of a marine terrace.
    C. expansion of the sandy beach.
    D. formation of sea stacks and pillars.

16. If you accidentally get caught in a rip current that is carrying you out to sea, what should you do?
    A. Swim parallel to the shore.
    B. Rest and float with the rip far out to sea and then swim back.
    C. Swim to shore because this is the shortest distance.
    D. Scream for help.
    Hint: Refer to Figure Story 17.13. Rip currents are rapidly moving backflows.

17. Coral atolls form in tropical oceans by the
    A. upward growth of coral from deep, submarine mountains.
    B. upward growth of coral on a continental shelf that is gradually subsiding.
    C. upward growth of coral on subsiding volcanic islands.
    D. upward growth of coral on exposed sections of the oceanic ridge system.
    Hint: Refer to Figure Story 8.16 and *Earth Policy* box 8.1.

18. What is the motion of individual water molecules as a wave travels?
    A. The molecules travel along with the wave.
    B. The molecules follow a straight up and down path.
    C. The molecules follow a roughly circular path.
    D. The molecules travel laterally along with longshore currents.

19. A result of wave refraction is that
    A. wave energy is concentrated on headlands and points.
    B. wave energy is concentrated in the bays.
    C. sediment is deposited in the vicinity of headlands, making them larger.
    D. wave energy is largely dissipated uniformly along the coastline.

20. Submarine canyons are formed by
    A. rifting of the lithosphere.
    B. erosion due to rivers.
    C. turbidity flows.
    D. glacial erosion.

21. On the diagram below, draw an arrow indicting the direction of the longshore current, which will be produced by the ocean waves hitting the beach.

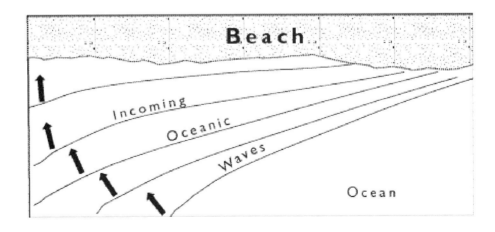

22. Sediments deposited on the ocean floor at depths greater than about 4 km are
    A. fine-grained dust (terrigenous sediments) and silica oozes.
    B. foraminiferal oozes.
    C. a mixture of foraminiferal and silica oozes.
    D. gas hydrates and carbonate sediments.

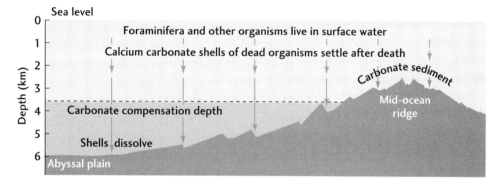

Figure 17.11. The carbonate-compensation depth is the level in an ocean below which calcium carbonate dissolves. As the shells of dead foraminifers and other carbonate-shelled organism settle into the deep waters, they enter an environment undersaturated in calcium carbonate and so dissolve.

CHAPTER 18

Landscapes: Tectonic and Climate Interaction

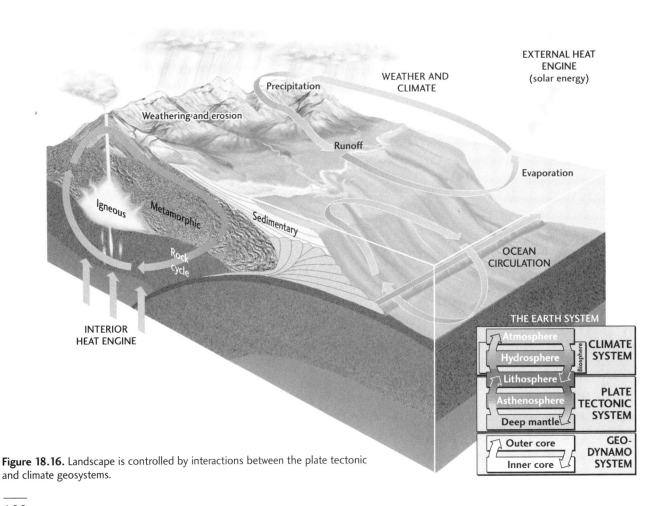

Figure 18.16. Landscape is controlled by interactions between the plate tectonic and climate geosystems.

Landscapes: Tectonic and Climate Interaction 181

Before Lecture

Before you attend the lecture, be sure to spend some time previewing the chapter. For an efficient preview, use the questions below.

Chapter Preview

- **What are the principal components of landscapes?**
  Brief answer: Landscapes are described in terms of their topography: elevation, the altitude of the surface of the Earth above sea level, and relief, the difference between the highest and the lowest spots in a region. Landscapes also consist of the varied landforms produced by geologic processes, such as erosion and sedimentation by rivers, glaciers, mass wasting, and wind.
- **How do the climate and plate tectonic systems interact to control landscape?**
  Brief answer: Tectonics affects the height and distribution of the crust and its composition. Climate affects weathering and erosion. Characteristics of the bedrock influence weathering and erosion rates.
- **How do landscapes evolve?**
  Brief answer: The evolution of landscapes depends strongly on the competition between uplift and erosion. For example, a landscape with high relief will form if tectonic activity is high, which in turn stimulates erosion. Erosion will at first enhance relief but over time water, wind, and ice will wear down the high spots and fill in the low spots with sediment.
- **Why don't mountains sink?**
  Brief answer: Like icebergs, mountains float. They float on Earth's mantle, which exerts a buoyant force and counters the force of gravity. During rapid erosion of mountain ranges summits may be uplifted to even greater heights because the mass of the mountain is reduced during erosion, resulting in isostatic uplift. Refer to Figure 18.17b.

Vital Information from Other Chapters

Congratulations! You have reached a point in your mastery of geology where you can really begin to make sense of a new landscape you may encounter during travel or while out on a hike. Chapter 18 on Landscapes draws heavily on all you have learned previously about tectonics and climate (note the chapter title). For that reason it will be even more important than usual to do a review in conjunction with your study of Chapter 18. Use the following quick reference list to expedite your review of important information from previous chapters. Note that this list is organized around the major theme of Chapter 18: *Landforms are shaped by the interaction of tectonic uplift and climate*.

Tectonics

How convergent margins produce uplift (Figure 2.9)

Landforms associated with faults and folds (Figures 11.10, 11.12, 11.14, 11.16, 11.17, 11.19, 11.21, and 11.22)

Plate tectonics and sedimentary basins (Figure 8.20)

Climate

Physical and chemical weathering (Figure 7.15)

How soil formation is driven by climate (Figure 7.16)

How mass wasting processes shape the landscape (Figure Story 12.6)

The relationship between mountains, atmospheric circulation, and deserts (Figures 13.3, 15.1, and 15.15)

Glacial landscapes (Figures 16.19 and 16.20)

Compared to ice and wind, running water plays the largest role in sculpturing the Earth's land surface today. A review of the hydrologic cycle (Figure 13.2) and the material in the last half of Chapter 14: *Streams: Transport to the Oceans* (Figures 14.14, 14.20, 14.21, and 14.22) will reinforce your understanding of the information presented in Chapter 18. This review will also serve as a great initial review in preparation for a final exam.

Web Site Preview

http://www.whfreeman.com/understandingearth
Identify North America's Landforms and *Understanding Landform Evolution* are Interactive Exercises definitely worth completing before or right after your first lecture on this topic.

During Lecture

One goal for lecture should be to leave class with a good set of answers to the preview questions.

- To avoid getting lost in details, keep the "big picture" in mind. Chapter 18 tells the story of how landscapes form. Essentially, "uplift proposes and erosion disposes."
- You might expect that a chapter on landscape forming processes would draw on virtually every previous chapter of the book and this is the case. A little review time before lecture (see above) will pay dividends.
- Chapter 18 provides a virtual tour of landforms created by geological processes. During lecture be alert to tips that will help you sort out links between a landscape and a certain set of geological circumstances.
- Even more important than the landscape features themselves are the processes that form each feature. You already know the processes. Your goal in this chapter is to understand how geologic processes work together to form a particular kind of landscape. Example: The Tibetan Plateau and the Himalayan Mountains are supported by buoyant continental crust, thickened during continental collision. Erosional unloading of the southern margin of the Tibetan Plateau may contribute to the height of the peaks in the Himalayan range.

After Lecture

The perfect time to review your notes is right after lecture. The checklist below contains both general review tips and specific suggestions for this chapter.

Check your notes: Have you...

- ☐ clearly described each landform (mountains, plateaus, Appalachian Valley and Ridge, etc.) and the geological processes that shape them. Use the text to check your notes. Consider adding simple sketches of each form.
- ☐ added visual material? Suggestions for Chapter 18. Make some simple sketches to help you learn the features that identify landforms in the chapter. Work some of these into a comparison chart, similar to Exercise 2: *Comparison of Some of the Landforms*. Tip: The answers for Exercise 2 (available at the end of this Study Guide) show some good examples of simple landform sketches.
- ☐ created a brief "big picture" overview of this lecture (using a sketch or written outline)? Hint: See Exercise 1: *Landscapes: Tectonic and Climate Interaction Flowchart*.

Study Tip: Learn by Drawing

Sketching simplified versions of landscape features into your notes is a helpful way to learn and remember because it activates both visual and kinesthetic learning modalities. Visual learners will remember material best after they look at and study a figure. Visual learners learn more if they enrich their notes with visual clues. For kinesthetic learners memory is activated by the act of drawing. So you learn as you look and draw.

Intensive Study Session

- **First, get the big picture in mind.** Take a look at Figure 18.16, which shows how tectonics and climate interact to produce landscape.
- **Next, review the main concepts of the chapter:** relief (Figure 18.3); development of ridges and valleys in folded mountains (Figure 18.10); erosion driven by the balance between stream power and sediment load (Figure Story 18.12); the dynamics of uplift (Figure 18.17); and classic models of landscape evolution (Figure 18.19). You will need to understand these figures to answer the exercise and review questions.

Study Tip for Figure 18.17 and Earth Issues box 18.1.

Study Figure 18.17 and *Earth Issues* box 18.1 as a package. Both relate uplift to climate change. They reinforce each other and make better sense together than they do when read separately.

- **Take the visual field tour of landforms.** Be sure you understand these landforms and know what they look like: aretes (Figure 18.6), mesas (Figure 18.8), valley and ridge topography (Figures 18.10 and 18.11), cuestas (Figure 18.14), and hogbacks (Figure 18.15).
- **Next, complete Exercises 1 and 2.** You will get the greatest return on your study time by working on these exercises because they will help you remember the most important ideas in the chapter.
- **Then, try the Review Questions.** Try answering each of the review questions to check your understanding of the lecture. Check your answers as you go, but do try to answer the question before you look at the answer. Pay attention to the test-taking tips provided. They will help you do better on your exams.
- **Sometime before your exam answer all 11 of the exercises at the end of this chapter.** These are short-answer questions and will not take long to complete if you know the material. Animations are provided on the web site for Exercises 3, 5, and 6.
- **Web Site Study Resources**
  http://www.whfreeman.com/understandingearth
  Complete the **Concept Self-Checker** and **Web Review Questions.** Pay particular attention to the explanations for the answers. Check the **Photo Gallery** for pictures of the landforms discussed in this chapter. The **Geology in Practice** exercise: *Can Erosion Make Mountains Higher?* is a good review and will reinforce the concepts presented in this chapter.

Exam Prep

Materials in this section are most useful during your preparation for quizzes and exams. The following **Chapter Summary** and **Practice Exercises and Review Questions** should simplify your chapter review. Read the **Chapter Summary** to begin your session. It provides a helpful overview that should refresh your memory.

Next, work on the **Practice Exercises and Review Questions.** In order to determine how you stand on mastery of this chapter, complete the exercises and questions just as you would a midterm. After you answer the questions score them. Finally, review any question that you missed. Identify and correct the misconception(s) that resulted in your answering the question incorrectly.

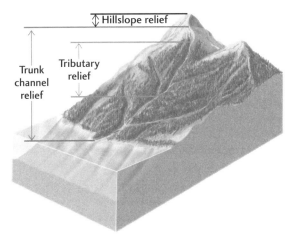

Figure 18.3. Relief is the difference between the highest and lowest elevations in a region. Three types of relief can be defined for a typical mountainous area. Hillslope relief is the decrease in elevation between mountain summits/ridgelines, and the point where channels begin. Tributary relief is the elevation decrease along the tributaries. Trunk channel relief is the decrease from the highest tributary to the end of the channel.

Chapter Summary

What are the principal components of landscapes?

- Landscapes are described in terms of their topography: **elevation,** the altitude of the surface of the Earth above sea level; **relief,** the difference between the highest and the lowest spots in a region; and the varied **landforms** produced by erosion and sedimentation by rivers, glaciers, mass wasting, and wind. Elevation is a balance between tectonic activity and erosion rate.

How do the climate and plate tectonic systems interact to control landscape?

- Tectonics (uplift and subsidence), erosion, climate, and the type of bedrock control the evolution of landscapes. Positive and negative feedbacks between tectonic process and climate dynamically adjust to change. Water, wind, and ice act to erode and transport rock material from the high spots and deposit it in the low spots. Refer to Figure 18.17 and *Earth Issues* box 18.1.

How do landscapes evolve?

- Formation of river valleys and bedrock erosion are controlled by a balance between stream power and sediment load. Refer to Figure Story 18.12.
- Landscapes go through different phases depending on tectonic activity and climate. For example, a landscape with high relief will form if tectonic activity is high, which in turn stimulates erosion. Erosion will at first enhance relief but over time water, wind, and ice will wear down the high spots and fill in the low spots with sediment.

- Current views of landscape evolution emphasize the dynamic equilibrium between erosion and tectonic uplift. Uplift competes with erosion. If uplift is faster, the mountain will rise; if erosion is faster, the mountains will be lowered. When tectonics dominates, mountains are high and steep, and they remain so as long as the balance is in favor of tectonics. When the rate of deformation wanes, higher erosion rates predominate, resulting in a gradual decrease in both relief and mean elevation.

Why don't mountains sink?

- Like an iceberg floating in water, most mountains are supported by a buoyant, low-density root of continental crust, which floats in the denser mantle. Isostatic rebound of thick continental crust in response to unloading due to differential erosion may cause mountain peaks to rise even higher.

The physical landscape is baffling in its ability to transcend whatever we would make of it.

—BARRY LOPEZ, *ARCTIC DREAMS*

Practice Exercises and Review Questions

Answers and explanations are provided at the end of this Study Guide.

Exercise 1: Landscapes: Tectonic and Climate Interaction Flowchart

To review the basic relationships between landscape relief, tectonic activity (uplift), and erosion, fill in the flowchart below. Place the following words in the correct positions on the flowchart:

tectonic activity
high relief
low relief
erosion
physical weathering
chemical weathering

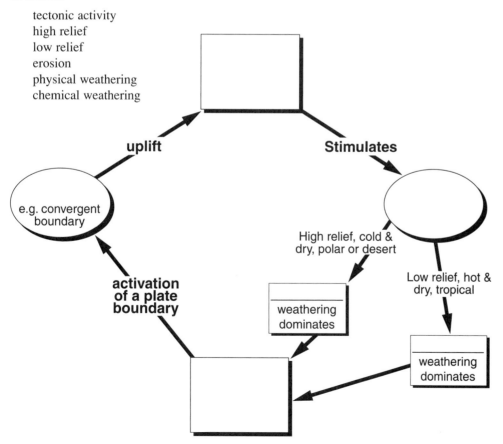

Exercise 2: Comparison of Some of the Landforms

Chapter 18 provided a virtual fieldtrip experience. To master the landforms discussed in this chapter it will be helpful to develop a comparison chart identifying the important features that distinguish each landform you learned about. Fill in the missing information under "Important Features." **Hint:** Use the text figures, captions, and accompanying text to help you. Then, make a very simple sketch of each landform. Sketching is both a thinking tool and a great kinesthetic learning tool. When you draw you tap the part of the brain that learns by moving.

Landform	Important feature(s)	Sketch (Hint: Keep it very simple)
Mesa See Figure 18.8.	A small plateau with _____ slopes on all sides. Held up by _____.	
Cuesta See Figure 18.14.	A structurally controlled cliff. Somewhat tilted beds with alternating weak and strong resistance to erosion. Typically undercut and asymmetrical.	
Hogback See Figure 18.15.	A structurally controlled cliff with beds that are _____. Ridge is more or less _____.	
Valley ridge topography See Figures 18.10 and 18.11.	In young mountains, upfolds (_____) form ridges and downfolds (_____) form valleys. As tectonic activity moderates and erosion digs deeper into the structures, the _____ may form valleys and syncline ridges.	

Review Questions

Answers and explanations are provided at the end of this Study Guide.

1. During the earliest stages of development of a river valley, the valley would have a
   A. simple V-shaped profile.
   B. simple U-shaped profile.
   C. low stream gradient.
   D. well-established floodplain.
   Hint: Refer to Figure Story 18.12.

2. Elevation is the result of
   A. tectonic activity.
   B. the balance between tectonic activity and erosion.
   C. erosion and deposition.
   D. deposition.

3. Relief is the
   A. difference between the highest and lowest point in a region.
   B. difference between the highest point and sea level.
   C. average height of a landscape.
   D. steepness of the slopes.

4. Which of the following are important controls on landscape evolution?
   A. Tectonics
   B. Climate
   C. Type of bedrock
   D. All of the above

5. The Earth has two fundamental levels on its surface. They are the
   A. land and oceans.
   B. crust and mantle.
   C. mountains and trenches.
   D. continental crust and ocean basins.

6. Mountain belts are found commonly in association with
   A. hot spots.
   B. convergent boundaries.
   C. transform faults.
   D. mid-ocean ridges.

7. An example of a relatively short-term positive-feedback process in landscape evolution is how
   A. a mountain peak may become higher as a result of erosion.
   B. uplift can cause erosion to slow down.
   C. rivers wash sediments out of subsiding basins.
   D. interactions at convergent boundaries result in low relief.
   Hint: Refer to Figure 18.17.

8. The Appalachian Valley and Ridge province is characterized by a landscape controlled by
   A. a series of regional faults that were active millions of years ago.
   B. glacial erosion and deposition since a continental ice sheet once covered the entire region.
   C. wind erosion and deposition partly constrained by zones of dense vegetation.
   D. an intricate series of anticlines and synclines.
   Hint: Refer to Figures 18.10 and 18.11.

9. Given that erosion by streams is controlled by a balance between stream power and sediment load, then in steep, wet terrain, stream power is
   A. high and sediment is transported away.
   B. low and sediment is transported away.
   C. high and sediment is deposited.
   D. low and sediment is deposited.
   Hint: Refer to Figure Story 18.12.

10. Cuestas and hogbacks are both long ridges of erosion-resistant rock. The difference between the cuesta and hogback is that
    A. much more steeply dipping beds form the cuesta.
    B. much more steeply dipping or vertical beds form the hogback.
    C. the cuesta tends to erode faster.
    D. hogbacks are very asymmetrical.

11. The debate about the dynamic interactions between uplift and climate is fueled by the observation that global cooling over the last few tens of millions of years coincides with the uplift of the Tibetan Plateau. One side of the debate argues that there was a **negative** feedback between uplift and climate because
    A. high erosion rates led to the removal of carbon dioxide—an important greenhouse gas—from the atmosphere which in turn led to further cooling, increased precipitation, and erosion.
    B. higher erosion rates resulted in higher rates of uplift due to isostatic adjustments.
    C. cooler climates allowed glaciers to grow on the mountain peaks and the weight of the ice isostatically depressed the mountains.
    D. none of the above.

12. Beryllium-10 is used as a method to date the age of river-terrace surfaces because beryllium-10
    A. is radioactive and its decay is determined by exposure to sunlight.
    B. is slowly leached out of rock once it is exposed at the surface.
    C. is slowly released from the rocks as they are uncovered by erosion.
    D. accumulates in the upper surface of the rock the longer it is exposed at the surface to cosmic ray bombardment.

CHAPTER 19

Earthquakes

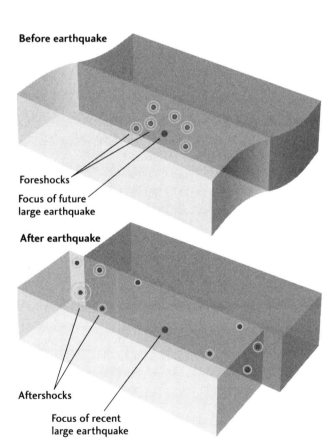

Figure 19.3. Foreshocks occur in the vicinity of, but before, the main shock. Aftershocks are smaller shocks that follow the main shock.

Before Lecture

Before you attend lecture be sure to spend some time previewing the chapter. For an efficient preview, use the questions below.

Chapter Preview

- **What is an earthquake?**
  Brief answer: An earthquake is a shaking of the ground caused by seismic waves that radiate out from a fault that moves suddenly. Elastic rebound explains why earthquakes occur. Refer to Figure Story 19.1.
- **What are the three types of seismic waves?**
  Brief answer: P (primary/compressional) waves, S (secondary/shear) waves, and surface waves. Refer to Figure Story 19.5.
- **What is earthquake magnitude and how is it measured?**
  Brief answer: Earthquake magnitude is a measure of the size of the earthquake. The Richter magnitude is determined from the amplitude of the ground motion. The moment magnitude is closely related to the amount of energy radiated by the earthquake. The Mercalli Intensity Scale is a more qualitative measure of the damage done by an earthquake.
- **Where do most earthquakes occur?**
  Brief answer: Most earthquakes occur along active plate tectonic boundaries but not all. Refer to Figure 19.12.

Vital Information from Other Chapters

A careful review of Chapter 11: *Folds, Faults, and Other Records of Rock Deformation* will provide you with important prerequisite information. Chapter 11 emphasizes brittle and plastic styles of rock deformation. Earthquakes are thought to be the result of the elastic behavior of solid rocks, analogous to the snap-back from a rubber band when it breaks. In particular, review pages 239–245, Figure Story 11.6, and Figure 11.11.

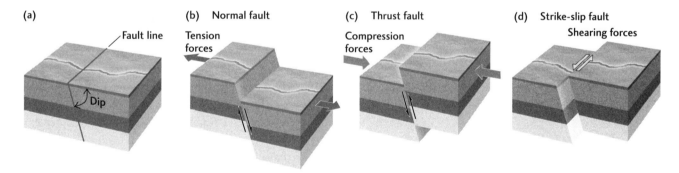

Figure 19.10. The three main types of fault movements that initiate earthquakes and the stresses that cause them: (a) situation before movement takes place; (b) normal fault due to tensile (tensional) stress; (c) thrust (reverse) fault due to compressive stress; (d) strike-slip fault due to shear stress.

Web Site Preview

http://www.whfreeman.com/understandingearth
Tectonic Forces in Rock Deformation Review for Chapter 11 and *Create an Earthquake* for Chapter 19 are **Online Review Exercises** worth completing before your first lecture on earthquakes.

During Lecture

One goal for lecture should be to leave class with a good set of answers to the preview questions.

- To avoid getting lost in details, keep the "big picture" in mind. Chapter 19 tells the story of earthquakes: how earthquake activity is measured, the seismic waves that are generated by earthquakes, and how earthquake activity is driven by plate tectonics (the location and characteristics of earthquakes is greatly influenced by the type of plate boundary).
- Focus on understanding the differences between the three kinds of seismic (P, S, surface) waves.

Note-Taking Tip: Marking Possible Test Items

As the end of semester approaches your instructor may mention that certain material will be covered on the end-of-semester exam. It will be helpful to have a systematic way of marking such comments so you won't miss them when you are studying your notes for the final. It's a great idea to adopt a standard abbreviation like **TQ** (Test Question) that you use to tag such comments. Obviously, it will save time if your **TQ** mark is very visible. Mark it in dark, large print and put it out in the margin where your eye will easily spot it during review. Formatting Tip: If you have adopted the strategy of leaving a column or the entire left page of your notebook blank for inserting visual material and special notes to yourself, that blank column will also be the perfect place to put your **TQ**. See the following example.

Example of How to Mark Your Notes for Possible Test Items

Blank column or page for visual material, TQ, and other additions to your notes.	Your notes:
	What is an Earthquake?
	Ground shakes
	Seismic wave from a moving fault → Ground shakes
TQ **(FS 19.5)** **Final May 10!**	What are the three types of seismic waves?
	P (primary) wave
	S (secondary) wave
	Surface waves

After Lecture

The perfect time to review your notes is right after lecture. The checklist below contains both general review tips and specific suggestions for this chapter.

Check your notes: Have you...

- ☐ identified the important points clearly? Hint: You should have headers in your notes for each of the questions in the *Before Lecture/Chapter Preview*.
- ☐ reviewed your notes right after class and filled in points that you didn't have time to record during lecture?
- ☐ marked possible test questions (TQ) in the margin? See example above.
- ☐ indicated important figures to study later?
- ☐ added visual material? Key visual material for Chapter 19 includes: Figure Story 19.1: *Elastic Rebound Theory of an Earthquake;* Figure Story 19.5: *Seismic Waves;* and the figure you will develop for Exercise 1 at the end of this Study Guide chapter. You can sketch simple versions of any aspect of these figures to help you remember the ideas.
- ☐ added a comparison chart to help you review key ideas? Suggestion: Complete Exercise 2: *Characteristics of Seismic Waves*. Add a copy of this exercise to your notes. It will be extremely useful for review purposes before the exam.

Intensive Study Session

Set priorities for studying this chapter. We recommend you give highest priority to activities that involve answering questions. We recommend the following strategy for learning this chapter.

- **Text.** Preview the key figures in the text, including Figure Story 19.1 (Elastic rebound theory of an earthquake), Figure Story 19.5 (Seismic waves), and Figure 19.6 (Determining the epicenter of an earthquake). You have to understand these figures to answer the **Practice Exercises and Review Questions.** Sometime before your exam answer all six of the exercise questions at the end of Chapter 19 in the text. These are short answers and won't take long if you know the material.
- **Practice Exercises and Review Questions.** Exercise 1: *Earthquake Focus vs. Epicenter* and Exercise 2: *Characteristics of Seismic Waves*. You will get the greatest return on your study time by working on these figures because they will help you remember the most important ideas in the chapter. Then, try the **Review Questions.** Try answering each of the review questions to check your understanding of the lecture. Check your answers as you go, but do try to answer the question before you look at the answer.
- **Web Site Resources**
  http://www.whfreeman.com/understandingearth
  Complete the **Concept Self-Checker** and **Web Review Questions.** Pay particular attention to the explanations for the answers. **Flashcards** will help you learn new terms. **The Online Review Exercises,** *Identify the Factors Contributing to Earthquakes at Plate Boundaries* and *Identify Factors Contributing to Earthquakes at Convergent Plate Boundaries*, are interactive exercises worth doing. Also, be sure to complete the **Geology in Practice** exercises to learn more about earthquakes.

Exam Prep

Materials in this section are most useful during your preparation for quizzes and exams. The following **Chapter Summary** and **Practice Exercises and Review Questions** should simplify your chapter review. Read the **Chapter Summary** to begin your session. It provides a helpful overview that should refresh your memory.

Next, work on the **Practice Exercises and Review Questions.** In order to determine how you stand on mastery of this chapter, complete the exercises and questions just as you would a midterm. After you answer the questions score them. Finally, review any question that you missed. Identify and correct the misconception(s) that resulted in your answering the question incorrectly.

Chapter Summary

What is an earthquake?

- An earthquake is a shaking of the ground caused by seismic waves that emanate from a fault that moves suddenly. When the fault moves, the strain built up over years of slow deformation by tectonic forces is released in a few moments as seismic waves.
- Elastic rebound theory explains why earthquakes occur. Over a period of time the application of stress causes rock to slowly deform (bend) elastically until its breaks and the rock snaps back as the fault moves. This is analogous to stretching a rubber band until it breaks and snaps back to sting your hand.

What determines the depth of an earthquake?

- The focus is a point along the fault from which the earthquake initiates.
- Earthquakes only occur in brittle rock, which can break and snap back elastically. At higher temperatures and confining pressures found at greater depths, rocks are ductile and do not break to generate earthquakes.
- Earthquakes occur within cold, brittle subducting oceanic lithosphere to a depth of about 700 km. Below this depth, the rock is too hot and soft to break.

Where do most earthquakes occur?

- Most earthquakes occur along crust plate boundaries but not all. Earthquakes at divergent plate boundaries are usually shallow, lower magnitude and a consequence of tensional stress. Convergent plate boundaries produce shallow and deep earthquakes of low to high magnitudes and commonly due to compressive stress. Transform faults produce shallow to moderately deep earthquakes of low to high magnitude and usually in response to shear stress. Refer to Figure 19.12.

What governs the type of faulting that occurs in an earthquake?

- The stress applied to the lithosphere is largely determined by the type of plate boundary. Tensional, compressional, and shear stresses determine the type of fault. Refer to Figures 19.10 and 19.11.

What is earthquake magnitude and how is it measured?

- Earthquake magnitude is a measure of the size of the earthquake. The Richter magnitude is determined from the amplitude of the ground motion. The movement magnitude is closely related to the amount of energy radiated by the earthquake. The Modified Mercalli Intensity Scale is a qualitative measure of the damage done by an earthquake.

What are the three types of seismic waves?

- There are three major types of seismic waves. Two types of waves travel through the Earth's interior: P (primary/compressional waves), which move through all forms of matter and move the fastest, and S (secondary/shear waves) waves, which move through solids only and move at about half the speed of P waves. The third type, surface waves, need a free surface like the

Earth's surface to ripple. They move more slowly than the interior waves but cause most of the destruction associated with earthquakes.

It often happens that the wave flees the place of its creation, while the water does not; like the waves made in a field of grain by the wind, where we see the waves running across the field, while the grain remains in place.
—LEONARDO DE VINCI

What causes the destructiveness of earthquakes?
- The destructiveness of an earthquake does not depend on the magnitude alone. In addition to ground motion, the duration of the earthquake, avalanches, fires, liquefaction, tsunamis, proximity to population centers, and the construction design of buildings all can amplify the destructiveness of an earthquake. Refer to *Earth Issue* box 19.1.

What can be done to mitigate the damage of earthquakes?
- The damage caused by earthquakes can be mitigated by: regulating the construction design of buildings in earthquake zones; bolting houses to their foundation; in your home, securing appliances connected to gas lines and tall furniture to walls and keeping heavy items at low levels; and having a community plan for dealing with emergencies generated by earthquakes. Geologists generate seismic-risk maps to aid public authorities with their evaluations. Refer to *Earth Issue* box 19.2.

Can scientists predict earthquakes?
- Scientists can characterize the degree of risk in a region, but they cannot consistently predict earthquakes.

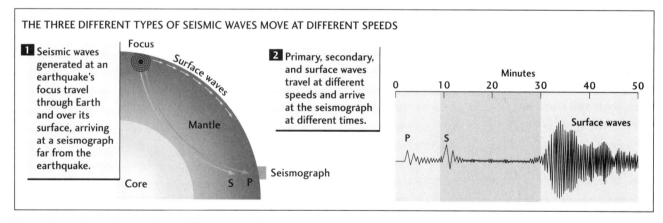

Figure Story 19.5. There are three different types of seismic waves that move at different speeds. (left) Seismic waves generated at an earthquake's focus travel through Earth and over its surface, arriving at a seismograph far from the earthquake. (right) Primary, secondary, and surface waves travel at different speeds and arrive at the seismograph at different times.

Practice Exercises and Review Questions

Answers and explanations are provided at the end of this Study Guide.

Practice Exercises

Exercise 1: Earthquake Focus vs. Epicenter

Label the diagram below by filling in the blanks using the following terms: fault scarp, earthquake focus, fault zone, earthquake epicenter. Use arrows to show the relative motion along the fault zone. The star marks the location from which fault movement propagated.

What type of fault is this? _____

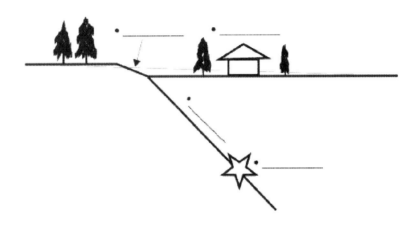

Exercise 2: Characteristics of Seismic Waves

Complete the following table by filling in the blank boxes. Figure 19.5 and the section of the textbook titled *Seismic Waves* will be helpful.

Seismic waves	P (primary) waves	S (secondary) waves	Surface waves
Relative speed		*Second fastest*	
Motion of material through which wave propagates			*Rolling/elliptical and sideways motions*
Medium through which wave will propagate			*Confined to the Earth's surface*
Analogy with common wave forms		*S waves propagation is difficult to visualize. It is somewhat analogous to the way cards in a deck of playing cards slide over each other as you shuffle the deck.*	

Exercise 3: Factors that Amplify the Damage Caused by an Earthquake

Briefly discuss five factors that can influence the damage caused by an earthquake.

1. _____
   _____

2. _____
   _____

3. _____

4. _____

5. _____

Review Questions

Answers and explanations are provided at the end of this Study Guide.

1. Elastic rebound theory says that earthquakes are produced when
   A. rocks deform plastically along a fault to produce anticlines and synclines.
   B. rocks abruptly slip past each other after an extended period during which elastic deformation is built up in the rocks.
   C. magma within the earth abruptly begins to flow and elastically deforms the surrounding rocks.
   D. abrupt movement along faults are caused by tidal forces.

2. The actual rupture-point within the crust that results in an earthquake is called
   A. the tsunami.
   B. the focus.
   C. the epicenter.
   D. static release.

3. The order of arrival of seismic waves at a recording station is
   A. P waves, S waves, surface waves.
   B. S waves, surface waves, P waves.
   C. P waves, surface waves, S waves.
   D. all waves arrive simultaneously.

4. In order to locate an earthquake epicenter, a minimum of _____ seismic stations is required.
   A. one
   B. two
   C. three
   D. between five and twelve, depending on the location of the earthquake
   Hint: Refer to Figure 19.6.

5. Richter Scale for earthquake magnitude measures the
   A. damage caused by the earthquake.
   B. amount of energy released by the earthquake.
   C. amount of ground motion.
   D. duration of the earthquake.

6. The ground motion generated by a Richter magnitude 8 earthquake is a factor of _____ times greater than a Richter magnitude 4 earthquake.
   A. 2   C. 1000
   B. 100   D. 10,000

7. Primary seismic waves (P waves), like sound waves,
   A. travel only through solid material.
   B. travel only through liquids and gas.
   C. travel through solid, liquid, or gas.
   D. are the slowest seismic waves.

8. Secondary seismic waves (S waves)
   A. travel parallel to P waves (parallel waves).
   B. can travel only through solid material.
   C. can travel through solid, liquid, or gas.
   D. are the fastest seismic waves.

9. A significant finding that supports the theory of plate tectonics is that most earthquakes occur
   A. randomly in the middle of lithospheric plates.
   B. at all active plate boundaries.
   C. only at plate boundaries that move toward each other.
   D. only at plate boundaries that slide past each other.

10. You just started a job as a county planner in Colorado when the Board of Supervisors mandates earthquake risk assessment. Your first task is to assess the potential for a major seismic event in an area that has experienced only a few minor earthquakes. So, you decide to
    A. install a state-of-the-art seismic recording station to monitor earthquake activity.
    B. develop a seismic risk map showing the likelihood of an earthquake based on the number that have occurred in the past.
    C. develop an earthquake protection plan with local and state officials.
    D. investigate the records for tsunamis.

11. Given four structures all built identically, which one would sustain the MOST damage during an earthquake when located the same distance from its epicenter?
    A. One built on a hillside composed of unfractured granite
    B. One built on a quartz-cemented sandstone formation
    C. One built on solid unfractured granite bedrock
    D. One built on a water-saturated stream delta
    Hint: Refer to the section titled *How Earthquakes Cause Damage* in your textbook.

12. Your seismograph has just recorded an earthquake. You are curious whether the earthquake occurred in North America or somewhere else in the world. Given the arrival time of 5 minutes for the first P waves and 10 minutes for the first S waves, you determine that the approximate distance between you and the earthquake is
    A. 100 kilometers.
    B. 1000 kilometers.
    C. 3000 kilometers.
    D. 7000 kilometers.
    Hint: Use the graph in Figure 19.6 to determine the distance between you and the earthquake.

13. Which of the following does NOT characterize a divergent plate boundary?
    A. Shallow-focus earthquakes
    B. Basalt eruptions
    C. Deep-focus earthquakes
    D. A rift valley
    Hint: Refer to Figure 19.13.

14. Of the states listed below, which one has the lowest potential for seismic hazard?
    A. Washington
    B. Utah
    C. Texas
    D. New York
    Hint: Refer to Figure 19.19.

15. The first motions of an earthquake as recorded from four different stations are displayed on the strike-slip fault shown below. Using this first motion data and Figure 19.11, determine the direction of relative motion along fault at the time of the earthquake.

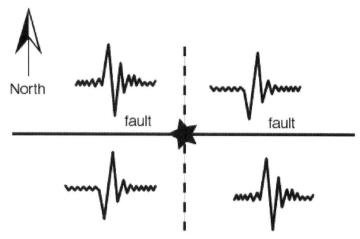

Bird's eye view of a fault zone with the first motion data for P waves arriving at four seismograph stations during an earthquake. The star marks the epicenter. The dashed line is a north-south reference line plotted perpendicular to the fault (solid line).

A. North side moved east (right) and south side moved west (left)
B. North side moved west (left) and south side moved east (right)
C. North side moved down and the south side moved up
D. North side moved up and the south side moved down

16. What type of fault produced the first motions shown in Question 15? _____

CHAPTER 20

Evolution of the Continents

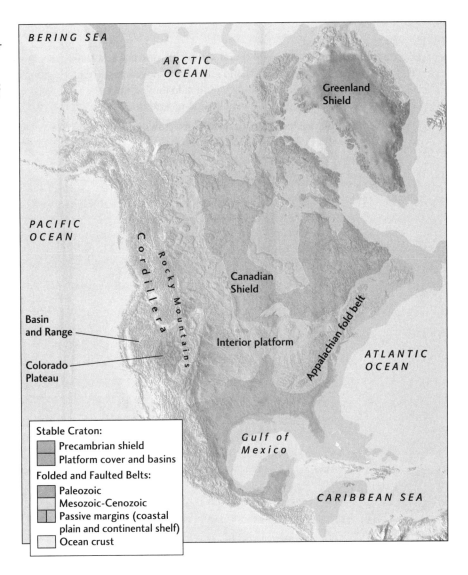

Figure 20.1. Major tectonic features of North America: Canadian Shield; interior platform; Cordilleran orogenic belt, including the Basin and Range; Colorado Plateau; Appalachian fold belt; coastal plain; and continental shelves of passive margins.

Before Lecture

Before you attend lecture be sure to spend some time previewing the chapter. For an efficient preview use the following questions.

Chapter Preview

- **What are the major geologic features of North America?**
  Brief answer: Review Figure 20.1. Pay particular attention to the location and characteristics of the active tectonic belt in western North America and the relatively stable continent of the ancient orogenic belts, platform, and shield. Look for patterns in age and location.
- **How do continents grow?**
  Brief answer: The silica-rich, iron-poor rocks in continents are produced mostly in subduction zones by magmatic differentiation and metamorphism of silica-rich sediments. Once produced, the continental rocks are difficult to subduct and recycle into the mantle because they are more buoyant than mantle material. Rifting and transform faulting typically breaks continents into small pieces while terrane accretion and continent collisions assemble pieces into larger continents. Refer to Figures 20.11 and 20.12.
- **How does orogeny modify continents?**
  Brief answer: As a styrofoam float resists being dragged under water, silica-rich, low-density continental crust is more buoyant than the mantle, and therefore continents resist being subducted. Instead, continental crust is damaged (deformed) and piled up (thickened) at convergent plate boundaries. Intense folding, faulting, detachment and thrust transport of sedimentary wedges and the formation of huge granitic batholiths thicken the overriding continental lithosphere. These processes can deform continental crust hundreds of kilometers from the convergence zone. Thicker continental crust tends to stand higher.
- **What is epeirogeny?**
  Brief answer: Epeirogeny is the gradual downward and upward movements of broad regions of the crust without significant folding or faulting. Heat, loading and unloading, and flow within the mantle may cause epeirogeny.
- **What is the Wilson cycle?**
  Brief answer: The Wilson cycle characterizes the sequence of events in the opening and closing of an ocean basin. A new ocean basin forms when a continent is rifted apart. As an ocean basin closes due to subduction of its ocean lithosphere, continents grow. A few times in Earth's history, the closure of a series of ocean basins resulted in the formation of a supercontinent, like Pangaea in the Early Permian. Refer to Figures 20.17 and 20.18.
- **How have the Archean cratons survived billions of years of plate tectonics?**
  Brief answer: Like giant sailboats, continents have cratonic keels that stabilize the raft of continental lithosphere against the effects of convective currents in the mantle and plate tectonic processes. It is hypothesized that the keels consisted of somewhat less dense mantle rocks that are about the same age as the Archean crust above them. Refer to Figure 20.24.

Study Tip

When confronted with new scientific terminology, sometimes the dictionary can help. For example, if you look up the terms *epeirogeny* and *orogeny* you would find the following origins for the roots of these terms:

      epeirogeny—Gk. *epeiros*, continent
      orogeny—Gk. *oros*, mountain

Vital Information from Other Chapters

To understand the evolution of the continents you will need to draw on diverse information from many previous chapters. While working on Chapter 20 keep the following concepts in mind and review them as much you need to:

Plate Tectonics

Divergent boundaries (Figure 2.6)

Convergent boundaries (Figure 2.9)

Forces that drive plate tectonics (Figure 2.16)

Igneous Rocks

Magma differentiation (Figure Story 5.5 and pages 95–98 in text)

Formation of igneous rocks at divergent and convergent plate boundaries (Figures 5.13 and 5.14 plus pages 102–107 in the text)

Rock Deformation

Folding and faulting (Figure Story 11.6 and Figure 11.11)

Thrust faulting (Figure 11.12)

Folding (Figure 11.16).

Appalachian fold belt (Figure 11.21)

Development of a fictitious geologic province (Figure 11.22)

Continental Margins

Characteristics of active and passive continental margins (Figure 17.8)

Landscapes

Erosion/Uplift feedback loop (Figure 18.17)

During Lecture

One goal for lecture should be to leave class with a good set of answers to the preview questions.

- To avoid getting lost in the details keep the "big picture" in mind. Chapter 20 examines the geologic history of the continents.
- Make a copy of Figure 20.18 and have it handy during lecture.

After Lecture

The perfect time to review your notes is right after lecture. The following checklist contains both general review tips and specific suggestions for this chapter.

Check your notes: Have you...

- ☐ added notes or sketches of material from previous chapters that you need to understand your lecture notes? This is a good idea for this chapter because it draws heavily on previous material. **Hint:** Refer to **Vital Information from Other Chapters.** There you will find a helpful list of material you may want to review or include as you rework your notes.
- ☐ created a brief "big picture" overview of this lecture? **Hint:** The Wilson Cycle provides a useful overview of this chapter. Refer to Figure 20.18 and consider adding sketches and text from that figure to your notes.

Intensive Study Session

Set priorities for studying this chapter. Try to give highest priority to activities that involve answering questions. We recommend the following strategy for learning this chapter.

- **Preview the key figures:** 20.1, 20.4, 20.8, 20.11, 20.12. 20.13, 20.15, 20.16, 20.18, 20.20, and 20.24. You will need to understand these figures to answer the **Review Questions.**
- **Complete Practice Exercises 1 and 2.** You will get the greatest return on your study time by working on these **Practice Exercises** because they will help you remember the most important ideas in this chapter.
- **Work in some review time as you study the evolution of the continents.** Reviewing is always a good idea and it is especially important for this chapter because it draws on so many ideas in previous chapters. Refer to **Vital Information from Other Chapters** for a helpful list of figures for review.
- **Try the Review Questions.** Try answering each of the **Review Questions** to check your understanding of the lecture. Check your answers as you go, but try to answer the questions before you look at the answers. Pay attention to the test-taking tips we provide. They will help you do better on your midterm.
- **Some time before your next exam, answer all the Exercises at the end of the chapter.** These are short answer and won't take long if you know the material. Note that helpful animations are provided on the Web site for Exercises 3, 5, and 6.
- **Web Site Study Resources**
  http://www.whfreeman.com/understandingearth
  Complete the **Concept Self-Checker** and **Web Review Questions.** Pay particular attention to the explanations for the answers. The **Geology in Practice Exercises** address the question: Do continents float on the mantle? **Flashcards** will help you to learn the new terminology in this chapter.

Exam Prep

Materials in this section are most useful during preparation for midterm and final examinations.

The following **Chapter Summary and Practice Exercises and Review Questions** should simplify your chapter review. Read the **Chapter Summary** to begin your session. It provides a helpful overview that should refresh your memory.

Next, work on the **Practice Exercises and Review Questions.** In order to determine how you stand on mastery of this chapter, complete the exercises and questions just as you would a midterm. After you answer the questions score them. Finally, review any question that you missed. Identify and correct the misconception(s) that resulted in your answering the question incorrectly.

Chapter Summary

What are the major geologic features of North America?

- Figure 20.1 shows the major tectonic features, including the Canadian Shield, interior platform, Cordillera and Appalachian mountain belts, coastal plain, and continental shelves.
- Figure 20.3 shows the sedimentary basins and domes of North America.
- The Appalachian fold belt and western Cordillera mountain belt are characterized in Figures 20.4, 20.5, and 20.11.

How do continents grow?

- Magmatic differentiation. Magmas generated in subduction zones tend to be more silica-rich because
  1. crustal material, like sediments, can be incorporated into the melt.
  2. partial melting and crystal settling differentiate the iron-rich and silica-rich materials.
- Continental accretion. Plate motions accrete buoyant (silica rich) rocks to continental margins by
  1. transfer of fragments from a subducting plate to a continental plate.
  2. closure of marginal basins to add thickened island-arc crust to the continent.
  3. lateral transport via strike-slip faulting along continental margins.
  4. suturing of two continental margins during their collision.
  Refer to Figures 20.11, 20.12, 20.15, 20.17, and 20.18 for illustrations and examples.

What are epeirogeny and orogeny?

- Epeirogeney refers to simple up and down movements without significant folding and faulting.
  Proposed mechanisms for vertical crustal movements are isostatic adjustments caused by loading or unloading of the crust, due to accumulation and melting of glacial ice, and heating or cooling the crust in association with continental rifting or a hot spot. Refer to Figure 20.20.
- Orogeny usually results from plate convergence and is characterized by severe deformation including extensive folding and folding. Examples of orogeny include the Appalachian and Rocky Mountains of the United States. But the most spectacular example on the planet is the Alpine-Himalayan belt. Refer to Figures 20.15 and 20.16 and the related section of text describing the formation of the Himalayas.

How does orogeny modify continents?

- As a styrofoam float resists being dragged under water, silica-rich, low-density continental crust is more buoyant than the mantle, and therefore continents resist being subducted. Instead, continental crust is damaged (deformed) and piled up (thickened) at convergent plate boundaries.
- Intense folding, faulting, detachment, and thrust transport of sedimentary wedges and the formation of huge granitic batholiths thicken the overriding continental lithosphere.
- These processes can deform continental crust hundreds of kilometers from the convergence zone. Thicker continental crust tends to stand higher.

What is the Wilson cycle?

- The Wilson cycle, named after J. Tuzo Wilson, is a flowchart that summarizes the general sequence of events in the evolution of continental crust. Refer to Figure 20.18.
- The cycle begins when the edges of continental cratons are rifted during the breakup of a supercontinent such as Pangea, then develop passive margins that collect ocean sediment, then become an active margin and orogen that experiences subduction and terraine accretion.
- A few times on Earth a series of orogenies produced supercontinents, like Rodina and Pangaea.

How have the Archean cratons survived billions of years of plate tectonics?

- The formation of silica-rich continental crust goes back at least 4.0 billion years. How have pieces of continental crust survived on the dynamic Earth for so long?
- Cratons (older and more stable parts of the continents) have keels that, like the hull of a boat, extend into the mantle.
- Cratonic keels appear to consist of somewhat lower-density rock, called peridotite.
- Keels appear to form at about the same time as the continental crust above them.
- It is hypothesized that the cold, strong, mantle keel more than 200 km thick helps to preserve the cratons from disruption by mantle convection and plate tectonic processes. Refer to Figure 20.24 below.

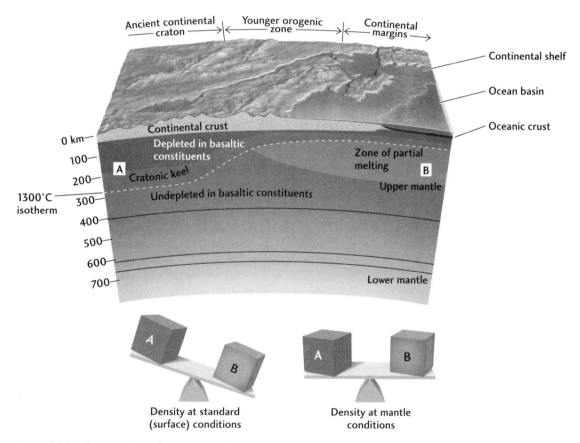

Figure 20.24 Cross section of a continent showing a cratonic keel extending to a depth of 250 km.

Practice Exercises and Review Questions

Answers and explanations are provided at the end of the Study Guide.

Practice Exercises

Exercise 1: Evolution of the Continents

Complete the sentences in the flowchart by filling in the blanks with words from the list. Refer to Figures 20.12, 20.15, and 20.18.

accretion and collision
continental plate
erosion
folding
hot
intrusion of magmas
magmatism
ocean basin
passive margin
stable craton
subducts
thick
thrust faulting

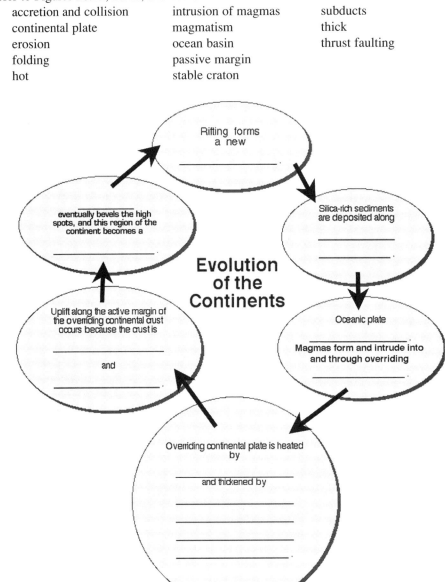

Exercise 2: Ocean Crust vs. Continental Crust.

Using information in Chapters 2, 5, 17, 18, and 20, complete the following table by filling in the blank boxes.

Characteristics	Ocean crust	Continental crust
Composition Rock type(s)		Very heterogeneous—can contain any rock but dominantly granitic and gneissic with a cover of sediments.
Density	3.0 g/cm^3	2.7 g/cm^3
Thickness	10 km	
Age		The ages of continental crust span 4 billion years.
Topographic features	Abyssal floor Ridge with axial rift Trenches Seamounts Hot spot island chains Plateaus	
Structure/Architecture	A model for the structure of the ocean crust is the ophiolite suite: deep-sea sediments, basaltic pillow lavas and dikes, and gabbro. (Note: peridotites are part of the mantle lithosphere, not the ocean crust.)	The architecture of the continents is complex. It consists of preexisting cratons, accreted microplates, island arcs, volcanic arcs, suture zones, ophiolite suites, and belts representing ancient orogenic zones. Sediments cover basement rock in the interior platform of the continent.
Origin		Orogenic processes and accretion of pre-existing crustal blocks along convergent plate boundaries.

Review Questions

Answers and explanations are provided at the end of the Study Guide.

1. The massive, interior regions of continents that have been stable for extensive periods of time are called
   A. mountains.
   B. plateaus.
   C. cratons.
   D. plains.

2. Intense deformation and accretion of continental crust occurs at
   A. convergent plate margins.
   B. transform margins.
   C. hot spots.
   D. divergent plate margins.

3. Which of the following is not associated with orogeny?
   A. Thrust faulting
   B. Intrusion of plutons
   C. Passive continental margin
   D. Metamorphism

4. The oldest rocks found on the continents are about
   A. 200 million years old.
   B. 1 billion years old.
   C. 4 billion years old.
   D. 4.5 billion years old.
   Hint: Refer to Figures 20.8 and 20.22.

5. Which of the following produces continental growth?
   A. Suturing of margins during a continental collision
   B. Transfer of crust during subduction
   C. Transfer of crust during strike-slip faulting
   D. All of the above

6. The Wilson cycle is
   A. a model for the evolution of landscapes.
   B. the sequence of events for the differentiation of magmas.
   C. the cycle of tectonic events related to continental evolution.
   D. a high-performance mountain bike developed in the Appalachian Mountains.

7. Much of the crust within the Cordillera of western North America _____ over the last 200 million years.
   A. accreted
   B. eroded
   C. subsided
   D. subducted

   Hint: Refer to Figures 20.11 and 20.12.

Top Ten Tips for Taking a Multiple-Choice Exam*

10. Answer the questions you know first. Mark items where you get stuck. Come back to harder questions later. Often you will find the answer you are looking for embedded in another, easier question.

9. Try to answer the item without looking at the options.

8. Eliminate the distracters. Treat each alternative as a true-false item. If false, eliminate it.

7. Use common sense. Reasoning is more reliable than memory.

6. Underline key words in the stem. This can be helpful when you are stuck. It may help you focus on what question is really being asked.

5. If two alternatives look similar, it is likely that one of them is correct.

4. Answer all questions. Unless points are being subtracted for wrong answers (rare) it pays to guess when you're not sure. Research indicates that items with the most words in the middle of the list are often the correct items. But be cautious. Your professor may have read the research too!

3. Do not change answers. Particularly when you are guessing, your first guess is often correct. Change answers only when you have a clear reason for doing so.

2. If the first item is correct, check the last. If the last item is "all (or none) of the above" you obviously need to read the other alternatives carefully. Missing an "all of the above" item is one of the most common errors on a multiple-choice exam. It is easy to read carelessly when you are anxious.

1. Read the directions before you begin!

---

* For optimal exam performance, review and use the **Final Exam Prep Worksheet** (see Appendix B: *Final Exam Prep* for details). The basic idea is to organize a systematic review of material that is divided into short study sessions.

---

8. Cratonic and other continental interior rocks are typically _____ those rocks found at active continental margins.
   A. much younger than
   B. older than
   C. younger than
   D. the same age as

   Hint: Refer to Figure 20.8.

9. Interior continental shields, or cratons, are traversed by ancient orogenic belts, but younger orogenic belts are different from ancient orogenic belts because
   A. their sediments have been subjected to extensive regional metamorphism.
   B. the crust is much thinner and cooler.
   C. they consist of hotter and thicker crust.
   D. they consist almost entirely of volcanic rocks.

10. A large portion of the Cordilleran orogenic belt may consist of
    A. accreted terranes.
    B. hot-spot volcanism.
    C. ophiolite suites.
    D. an ancient craton.
    Hint: Refer to Figure 20.11.

11. Orogeny is taking place today
    A. along east coast of North America.
    B. along west coast of South America.
    C. in the center of North America.
    D. in the Canadian Shield.

12. Which of the following statements about orogenic systems is NOT valid?
    A. Orogeny is initiated by rifting as extension begins to open up a new ocean basin.
    B. Orogeny is initiated by subduction and the evolution of an active convergent margin.
    C. Large volumes of granite are intruded during orogenies.
    D. Folding and thrusting of preexisting rocks contributes to crustal thickening in the orogen.

13. The _____ were produced by convergent plate boundary processes, including collision, during the Paleozoic.
    A. Cascade Mountains of northwestern North America
    B. Rocky Mountains of Colorado
    C. Appalachian Mountains along the eastern margin of North America
    D. Cordillera of western North, Central, and South America
    Hint: Refer to Figures 20.4 and 20.17.

14. The Cordilleran orogeny was initiated by formation of
    A. a continental rift in western North America.
    B. a subduction zone and convergent plate boundary along the western edge of North America.
    C. an impact of a comet-size object in the Pacific Ocean adjacent to the western edge of North America.
    D. a line of mantle or hot spots which causes doming and rifting through out western North America.

15. The western Cordillera of North America is topographically higher than the Appalachians mainly because
    A. orogeny occurred more recently in the Cordillera, so the crust is still thicker.
    B. granite batholiths were intruded in the Cordilleran belt.
    C. the Appalachians eroded faster because they consist mostly of soft sedimentary rocks.
    D. the Appalachians never reach the elevation of the Cordillera because collisions do not generate high mountains.

16. Your assignment as a geologist is to map out ancient orogenic zones in what is now the stable interior of a continent. Recent epeirogenic uplift has resulted in good exposures of the ancient basement rocks. Which of the following features would NOT be evidence of an ancient orogenic zone?
    A. Intensely deformed sedimentary rocks
    B. Lava flows and thick layers of volcanic tuff
    C. Many granitic plutons that are all about the same radiometric age
    D. Widespread and relatively thick accumulations of coral-rich limestone, sandstones, and shales

17. Why are the Sierra Nevada Mountains so much higher than the continental surfaces west and east of them?
    A. The crust is proably thicker and/or hotter beneath the Sierra Nevada.
    B. The crust beneath the Sierra Nevada must be very thin and hot.
    C. The crust beneath the Sierra Nevada is probably more dense than that to the east and west.
    D. The Sierra Nevadas are most probably part of an ancient spreading center that is no longer active.

18. Mountains are both the source and product of sediments because
    A. most sediments are subducted with the ocean lithosphere and thereby contribute to subduction zone magmatism.
    B. most sediments shed off mountains end up on the margin of a continent and are eventually accreted to the edge of continental crust by orogenic processes associated with convergent plate margins.
    C. the melting of sediments results in mafic igneous rocks characteristic of continental crust.
    D. most sediments end up on the deep ocean floor where they sit for billions of years.

19. In the context of plate tectonics, a reasonable sequence of events for an orogenic "cycle" is
    A. hot spots → subduction → rifting → orogeny.
    B. rifting → passive margin → subduction → orogeny → uplift.
    C. rifting → collision → subsidence → erosion → uplift.
    D. transform faulting → uplift → volcanism → orogeny.
    Hint: Refer to Figure 20.18.

20. The Earth's oldest continental crust can be found
    A. in active orogenic zones.
    B. on the ocean floor.
    C. in continental shield regions.
    D. along the margins of the continents.

21. The growth of continents occurs at
    A. hot spots
    B. subduction zones
    C. rift zones
    D. transform faults

22. Where could you go today on Earth to find an orogenic system with strong similarities to the Cordillera of North America? The study of this active orogenic system would provide you with a better understanding of the geologic history of western North America.
    A. Andes Mountains.
    B. Himalayas.
    C. Appalachians.
    D. East Africa Rift.

23. Epeirogeny is associated with
    A. subduction zones where convergence causes rapid vertical uplift.
    B. stable interior platforms within continents where isostatic adjustments result in gradual uplift or subsidence.
    C. continental collisions where the highest mountains form.
    D. the Wilson cycle for the evolution of continents.
    Hint: Refer to Figure 20.20.

24. Continental cratons seem relatively immune from deformation by plate tectonic processes because
    A. of deep, strong, mantle keels beneath them.
    B. they are always located at the center of the continent far from active plate boundaries.
    C. the continental crust is very strong.
    D. they are covered with sediments.
    Hint: Refer to Figure 20.24.

25. A modern example of how crust can be transported laterally along a continental margin is the
    A. amalgamation of the southwestern Pacific islands.
    B. mid-Atlantic Ridge.
    C. Appalachian orogeny.
    D. San Andreas strike-slip fault.
    Hint: Refer to Figure 20.6.

CHAPTER 21

Exploring Earth's Interior

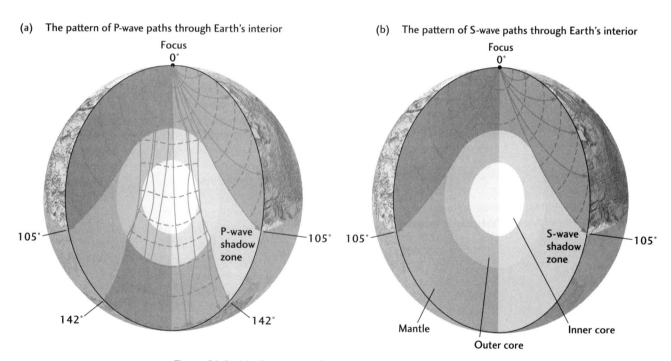

Figure 21.2. (a) The pattern of P-wave paths through Earth's interior. Lines show the progress of wave fronts through the interior at 2-minute intervals. The P-wave shadow zone extends from 105° to 142°. P waves cannot reach the surface within this zone because of the way they are bent when they enter and leave the core. (b) The larger S-wave shadow zone extends from 105° to 180°. Although S waves strike the core, they cannot travel through its fluid outer region and therefore never emerge beyond 105° from the focus.

Before Lecture

Previewing will greatly increase your understanding of the lecture. For an efficient preview use the following questions.

Chapter Preview

- **What do seismic waves reveal about the Earth's interior?**
  Brief answer: Seismic waves reveal that the Earth has a concentrically zoned internal structure. The felsic crust lies on a denser ultramafic mantle composed mostly of peridotite. The crust and upper mantle make up the rigid lithosphere. Beneath the lithosphere lies the asthenosphere, the weak layer of the mantle across which the lithosphere slides in plate tectonics. The liquid outer core and solid inner core are mostly iron. Refer to Figures 21.6 and 21.7.
- **What has seismic tomography revealed about structures in the mantle?**
  Tomographic images show how tectonic plates very from very thin under the mid-ocean ridges to very thick under continental cratons. Many features of mantle convection are also revealed. Refer to Figure 21.10.
- **How hot does it get in Earth's Interior?**
  Refer to Figure 21.9.
- **What do Earth's gravity field and isostatic rebound tell us about the interior?**
  Brief answer: The observed gravity field is in agreement with the pattern of mantle convection inferred from seismic tomography. Measuring the rate of post-glacial isostatic rebound provides information on the viscosity of the mantle and how it affects rates of uplift and subsidence of the buoyant lithosphere. Refer to *Earth Issues* box 21.1.
- **What does Earth's magnetic field tell us about the fluid outer core?**
  Brief answer: The Earth's magnetic field is produced by convective motions of electrically conducting iron-rich fluid in the outer core.
- **What is paleomagnetism and what is its importance?**
  Brief answer: The Earth's magnetic field flips back and forth over geologic time. Preserved in some rocks is a record of past changes in the orientation of Earth's magnetic field. Refer to Figure 21.15.

Vital Information from Other Chapters

It is very important to review the information on seismic waves presented in the section *Studying Earthquakes* at the beginning of Chapter 19. Pay particular attention to the subsection *Seismic Waves* and Figure Story 19.5 and be sure you understand the distinctions between P-waves and S-waves. A quick review of models for mantle convection will also be helpful. The key information is covered in Figures 1.11, 2.17, and 6.22. Finally, take another look at Figure Story 2.11, which was your first exposure to paleomagnetism in the text.

During Lecture

One goal for lecture should be to leave class with a good set of answers to the preview questions.

- To avoid getting lost in details keep the "big picture" in mind: Chapter 21 tells the story of the interior of the earth, its structure and composition, and how Earth's interior supplies heat energy to drive geological processes. Key points:
  - Earth's interior is a concentrically zoned structure.
  - Continents float on the mantle.

- Mantle behaves like a viscous fluid.
- P & S waves reveal a liquid outer core and solid inner core.
- Heat transfer occurs via convection.
- Earth's magnetic field is best understood as a **geodynamo:** convective movement (driven by Earth's internal heat) generates an electromagnetic field.
• Focus on understanding Figures 21.2, 21.5, 21.7 and 21.9. Hopefully, you will have looked at these before coming to lecture. That way it will be easy to follow the lecture: you can simply annotate the figures with important new material provided by your instructor and underline material in the captions.

Note-Taking Tip

We all have moments when we don't understand a point being made in lecture. But even when you are momentarily confused, it is important to continue taking notes. Hopefully, the necessary insight will come to you. If it does not, the notes you take will provide a clue to what you need to investigate further in your text or in a conversation with your instructor.

After Lecture

Check your notes: Have you...

☐ annotated figures in the text with important material discussed by your instructor? Suggestions for Chapter 21: Figures 21.2, 21.5, 21.7, and 21.9.
☐ added visual material? Because this chapter depends heavily on material from Chapters 1, 2, and 19, it may be useful to quickly sketch key ideas about P- and S-waves and mantle convection in your notes. (See **Vital Information from Other Chapters** for suggested material.)
☐ added a brief "big picture" overview of this lecture in your own words?

Intensive Study Session

Set priorities for studying this chapter. We recommend giving highest priority to activities that involve answering questions. We recommend the following strategy for learning this chapter.

- **Preview the key figures** 21.2, 21.5, 21.7, 21.9, 21.10, 21.11, 21.15, and *Earth Issues* box 21.2. You have to understand these figures to answer the **Review Questions.**
- **Complete Practice Exercises 1 and 2.** You will get the greatest return on your study time by working on these exercises because they will help you remember the most important ideas in the chapter.
- **Try the Review Questions.** Try answering each of the review questions to check your understanding of the lecture. Check your answers as you go, but do try to answer the question before you look at the answer. Notice the test taking tips that are interspersed with these questions. They are designed to help you do better on your next exam.
- **Some time before your next exam answer all the Exercises at the end of the chapter.** These are short answer and won't take long if you know the material. Notice that helpful animations are provided on the Web site for some of the chapter exercises.

- **Web Site Study Resources**
  http://www.whfreeman.com/understandingearth
  Complete the **Concept Self-Checker** and **Web Review Questions.** Pay particular attention to the explanations for the answers. Did you know that Cleopatra adored peridot, a gemstone from the upper mantle? Check out **Geology in Practice** to find out more about "Cleopatra's Emeralds" (peridot) and the composition of the upper ultramafic mantle. **Flashcards** will help you review the new terminology in this chapter.

Exam Prep

Materials in this section are most useful during preparation for exams.

> ### Exam Prep Tip: Get Organized for Finals Week
> The end of semester is approaching. It's time to get organized for taking exams in all your courses. Take a look at the **Final Exam Prep Worksheet** (Appendix B at the end of this Study Guide). There you will find many useful ideas about how to be successful as you enter the home stretch.

The following **Chapter Summary** and **Practice Exercises and Review Questions** should simplify your chapter review. Read the **Chapter Summary** to begin your session. It provides a helpful overview that should refresh your memory.

Next, work on the **Practice Exercises and Review Questions.** Complete the exercises and questions just as you would for an exam to see how you stand in regard to mastery of this chapter. After you answer the questions, score them. Finally and most important of all, review any question that you missed. Identify and correct the misconception(s) that resulted in your answering the question incorrectly.

Chapter Summary

What do seismic waves reveal about the layering of Earth's crust and mantle?

- Seismic waves reveal that the Earth has a concentrically zoned internal structure. Felsic continental and mafic ocean crusts lie on a denser ultramafic mantle consisting of iron-rich silicates, like peridotite. Refer to Figure 21.7
- The Moho or Mohorovicic discontinuity in seismic wave velocities marks the boundary between the crust and the mantle. Refer to Figures 21.5 and 21.6.
- Earth's tectonic plates are large fragments of the lithosphere, which includes the crust and uppermost, rigid mantle. Below the lithosphere in the upper mantle is a weak (soft) zone called the asthenosphere. Refer to Figure 21.7.
- Abrupt increases in seismic wave velocities coupled with laboratory studies on high-pressure minerals suggest that there are zones at progressively greater depths within the mantle where the crystal structures of minerals collapse (change phase) under the intense pressure to form more compact atomic structures and therefore different minerals.

What do seismic waves reveal about the layering of Earth's core?

- P wave and S-wave shadow zones reveal a liquid outer core and a solid inner core. Refer to Figure 21.2.
- P wave velocities in the core, the natural abundance of iron in nature, the existence of iron-nickel meteorites, the Earth's strong magnetic field, and the need

for a very dense core to account for the overall mass of the Earth all support an iron-nickel composition for the Earth's core.

What has seismic tomography revealed about structures in the mantle?

- These CAT scan–like images of the Earth's interior reveal how tectonic plates vary from very thin under the mid-ocean ridges to very thick under continental cratons. Seismic tomography also reveals features, like sinking slabs of lithosphere and superplumes, associated with mantle convection. Refer to Figure 21.10.

How hot does it get in Earth's interior?

- Within normal continental crust, temperature increases at a rate of 20° to 30°C per kilometer. The rate of temperature increase slows way down in the mantle and core. The most rapid change in temperature (steepest geothermal gradient) occurs in the outermost layer of our planet. This is not surprising, if you consider how rapidly temperature changes from the outside to the inside of a kitchen oven door. Refer to Figure 21.9.
- Seismic and laboratory studies suggest that the temperature in the outer liquid core is higher than 3000°C and at the Earth's center the temperature may reach 6000° to 8000°C. Refer to Figure 21.9.

What does Earth's gravity and isostatic rebound tell us about the interior?

- Measuring the rate of post-glacial isostatic rebound provides information on the viscosity of the mantle and how it affects rates of uplift and subsidence of the buoyant lithosphere. Refer to *Earth Issues* box 21.1.
- The observed gravity field is in agreement with the pattern of mantle convection inferred from seismic tomography. *Earth Issues* box 21.1

What does Earth's magnetic field tell us about the fluid outer core?

- Earth's magnetic field basically looks like a dipolar bar magnet. Refer to Figure 21.11.
- A geodynamo explains how Earth's magnetic field is generated. Rapid convection in the molten outer core is thought to stir up electrical currents in the conducting iron to create the major component of the magnetic field.
- The magnetic field changes strength, position, and polarity over time. There is a nondipole component. Both the dipole and nondipole components exhibit secular variations over time scales of decades. Reversals in polarity also occur over time spans of thousands of years. Refer to Figure 21.15.

What is paleomagnetism and what is its importance?

- Preserved in some rocks is a clear record of past changes in the orientation of Earth's magnetic field. The chronology of magnetic field reversals has been worked out so that the direction of remnant magnetization of a rock formation is often an indicator of its stratigraphic age. Refer to Figures 21.14 and 21.15.

    Note: The pattern of magnetic anomalies produced by paleomagnetic reversals recorded by ocean floor rock provided important evidence for seafloor spreading. Refer to Figure Story 2.11.

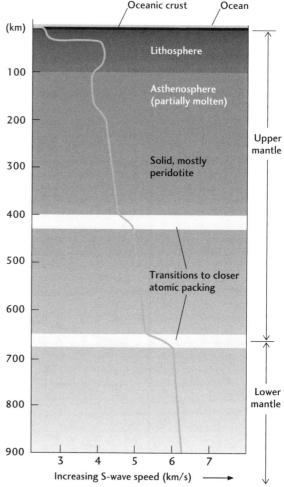

Figure 21.7. The structure of the mantle showing the S-wave velocity to a depth of 900 km. Changes in velocity mark the strong lithosphere, the weak asthenosphere, and two zones in which changes occur because of increasing pressure forces a rearrangement of the atoms into denser and more compact crystalline structures. [After D. P. Mckenzie, "The Earth's Mantle." *Scientific American* (September 1983): 66.]

Practice Exercises and Review Questions

Answers and explanations are provided at the end of the Study Guide.

Practice Exercises

Exercise 1: The Earth's Interior Layers

Fill in the blanks. Use Figures 21.5, 21.6, and 21.7 as references, in addition to the section *The Layering and Composition of the Interior* in this chapter.

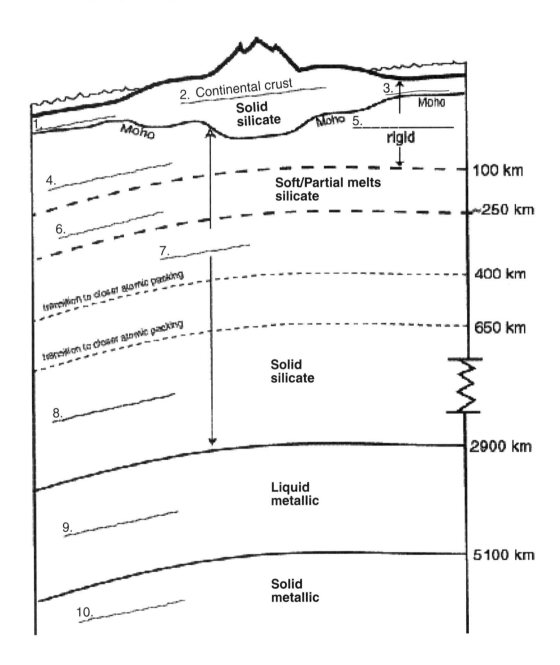

Exercise 2: The Characteristics of Earth's Internal Layers

Complete the table by filling in the blanks and completing the sentences. Shaded boxes remain blank. Refer to the text and figures in Chapters 1, 2, 5, 20, and 21.

Layers	Volume % of total	Mass % of total	Density g/cm³	Physical state	Composition	Observations and evidence that support the characteristics of the layer
Crust	**0.60**	**0.42**				Very heterogeneous. 40–65 km thick. Formed at convergent boundaries by orogenic processes.
Continental crust	0.44	0.25	≈ 2.7	Solid		
Ocean crust	0.16	0.17	≈ 3.0			Very homogeneous. About 10 km thick. Formed at _____ from _____ mantle rocks.
Mantle	**83.02**	**67.77**	3.3–5.7			
Mantle lithosphere						Crust and mantle lithosphere makes up Earth's _____. Thickness ranges from 0 km at spreading centers to 200 km beneath continents. S-wave velocities _____ through it.
Asthenosphere						Weak zone. A _____ melting. Reaches close to the surface at spreading centers and deepens under older seafloor. S-waves _____ and are partially absorbed.
Lower mantle						Abrupt _____ in S-wave velocities at 400 and ___ km mark changes in mantle structure—collapse of the _____ of minerals.
Core		**31.79**				
Outer core	15.68		9.9–12.2			P-waves slow down; S-waves are _____. Iron-nickel composition is consistent with bulk density of the _____.
Inner core	0.70		12.6–13.0			P waves suddenly _____ at 5100 km. S-waves are _____ through inner core. Composition also consistent with the natural abundance of iron and meteorites.

Review Questions

Answers and explanations are provided at the end of this Study Guide.

1. Earth's core has a radius that is about _____ of the Earth's radius.
   - A. 1/8
   - B. 1/4
   - C. 1/2
   - D. 3/4

   Hint: Refer to Figure 21.5.

2. The thickness of Earth's tectonic plates is
   - A. the same on the continents as under the oceans.
   - B. at its thinnest under the oceans.
   - C. at its thinnest in the continents.
   - D. completely unknown.

   Hint: Refer to Figure 21.6 and 21.7.

3. The likely composition of the upper mantle is
   - A. felsic.
   - B. mafic.
   - C. ultramafic.
   - D. carbon (diamonds).

4. Which of the following constitutes the rigid, outer layer of Earth's tectonic plates?
   - A. Asthenosphere
   - B. Lithosphere
   - C. Crust
   - D. Mantle

5. Continental crust has an overall composition corresponding closely to that of
   - A. ultramafic.
   - B. mafic.
   - C. felsic to intermediate.
   - D. peridotite.

6. The lithosphere is a _____ layer, as opposed to the asthenosphere.
   - A. plastic
   - B. fluid
   - C. weak
   - D. rigid

7. The inference that Earth's outer core is liquid is supported by the observation that
   - A. P-waves do not pass through it.
   - B. S-waves do not pass through it.
   - C. P-waves travel more rapidly through it.
   - D. S-waves travel more slowly through it.

8. The highest-density component of the Earth is
   - A. the mantle.
   - B. continental crust.
   - C. the core.
   - D. the whole Earth.

9. Earth's north magnetic pole is located
   - A. at the north geographic pole.
   - B. in Alaska.
   - C. between Greenland and Baffin Island.
   - D. in China.

   Hint: Refer to an atlas.

10. The Earth's magnetic field is thought to be generated by
    - A. permanent magnetism of minerals within the mantle.
    - B. permanent magnetism of the solid iron-rich inner core.
    - C. electrical currents generated by movement of the liquid outer core.
    - D. electrical currents generated by convection in the asthenosphere.

11. The asthenosphere is a _____ layer, as opposed to the lithosphere.
    - A. brittle
    - B. weak
    - C. molten
    - D. rigid

Test-Taking Tips: Test Taking and Learning Style

Knowing your learning style can help during exams. Consider the following.

Visual Learners
- If you are a visual learner, you probably pay better attention to directions that are written out (visual) than to spoken directions. Rely on written directions when they are available. If your exam proctor gives the directions verbally without a visual you must compensate. Make yourself listen and don't hesitate to ask as many questions as you need to to get the directions straight.
- When you get stuck on an item, activate your visual memory. Close your eyes and picture flowcharts, pictures, field experiences, or even lines of text.

Auditory Learners
- You probably pay better attention to directions that are spoken than to directions that are written out. Rely on spoken directions when they are available. If the directions are on a slide, compensate. Make yourself read them.
- Repeat written directions quietly to yourself (moving your lips is often enough)
- When you get stuck, remember your lecturer's voice covering this section.

Kinesthetic Learners
- You probably do best with directions that allow you to work an example. Unfortunately, it is a rare classroom exam that provides examples or samples as part of the directions, so you will need to make up your own. Take a minute to translate the directions into something you can do, or ask the instructor for a sample or example. Be sure to interact. Remember, you learn by doing. Kinesthetic learners find a variety of things helpful when they get stuck on a test item. Try some of these:
- When you get stuck move in your chair or tap your foot, to trigger memory.
- Feel yourself doing a lab procedure.
- Sketch a flowchart to unlock memory of a process.
- Stuck on a geology problem? Sketching what is being described (what is "given" in the problem) may unlock your memory and get you started.

12. Mineral grains in sediments become magnetized by the Earth's magnetic field when
    A. they are struck a sharp blow by a meteorite.
    B. iron-rich minerals align parallel to the Earth's magnetic field.
    C. the Earth's magnetic field reverses itself.
    D. electricity from lightning strikes passes through the lava beds.
    Hint: Refer to Figure 21.14.

13. Which layer of the Earth experiences the most rapid increase in temperature with increasing depth?
    A. Lithosphere
    B. Asthenosphere
    C. Mantle
    D. Liquid outer core
    Hint: Refer to Figure 21.9.

14. Between which boundary in the Earth's interior does the greatest change in composition occur?
    A. Lithosphere-asthenosphere boundary
    B. Crust–mantle boundary
    C. Mantle–core boundary
    D. Boundary between the outer and inner core

15. The crust is typically thickest beneath
    A. high mountain ranges and plateaus on the continents.
    B. ocean spreading centers.
    C. continental interiors like the Great Plains in North America.
    D. passive margins of continents where topography is very flat.
    Hint: Refer to Figure 21.6.

16. Significant increases in S-wave velocity at about 400 and 650 kilometers depth are explained by
    A. changes in the chemical composition of the mantle.
    B. a collapse of the crystal structures to more close packed forms.
    C. changes in the degree of partial melting within the mantle.
    D. a rapid increase in temperature.
    Hint: Refer to Figure 21.7

17. It is believed that the Earth's core is composed mostly of iron because
    A. iron is naturally very abundant.
    B. most meteorites representing the interstellar matter from which the planets formed are rich in iron.
    C. iron is very dense; its presence in the core would account for the Earth's average density.
    D. all of the above.

18. Without any knowledge of what seismic waves tell us about the Earth's interior, why is it unreasonable to assume that a large portion of the lower mantle is molten?
    A. Actually, it could be molten. We just don't see evidence for it because silicate magmas are trapped within the Earth due to the confining pressures.
    B. Direct measurements show that the temperature at the core/mantle boundary is not high enough to melt the lower mantle.
    C. The magnetic field strength would be greatly reduced if more of the Earth's interior was molten.
    D. Silicate magmas are less dense and would rise to the surface, so we should observe widespread volcanic activity across the Earth's surface. The molten iron-rich outer core is too dense to rise.

19. As rocks experience increased pressure with depth, P-waves in general _____ as they migrate through them.
    A. travel faster. C. travel at the same velocity.
    B. travel slower. D. rapidly die out.

20. When a reversal of the Earth's magnetic field occurs, the
    A. sense of rotation of the Earth is also reversed.
    B. Earth flips over in its orbit.
    C. magnetic polarity of the Earth reverses, such that the north end of a magnetic compass needle points toward the south geographic pole.
    D. almost all the igneous sedimentary rocks of the ocean floor reverse in magnetization to match the new orientation of the magnetic field.

21. The Moho, or Mohorovicic discontinuity between the crust and the mantle, was first detected based on
    A. the abrupt decrease in seismic velocities as they cross it.
    B. the abrupt increase in seismic velocities as they cross it.
    C. the S-wave shadow zone through which S-waves do not pass.
    D. the observation that no earthquakes occur below the Moho.

22. Supporting evidence for heat transfer by convection within the mantle comes from
    A. tomography and Earth's gravity field.
    B. the bulk density of the Earth.
    C. post-glacial isostatic rebound.
    D. Earth's magnetic field.

23. The "Cretaceous quiet zone" refers to
    A. a period when dinosaurs were very sedate.
    B. a time when plate motions slowed way down.
    C. a break in volcanic activity on Earth due to a lack of mantle superplume activity.
    D. an especially long period of normal polarity of Earth's magnetic field.
    Hint: Refer to Figure 21.15.

24. The P-wave shadow zone is caused by the way the Earth's core
    A. refracts the seismic waves.
    B. reflects the seismic waves.
    C. absorbs the P-waves.
    D. blocks the P-waves.

25. As the lithosphere cools slowly by the conduction of heat, it becomes
    A. less dense and rises.
    B. soft and weak.
    C. denser and rises.
    D. denser and subsides.
    Hint: Refer to Figure 21.8.

CHAPTER 22

Energy and Material Resources from the Earth

A hard fact of life is that an equal per capita sharing of the world's available reserves would not be enough to bring everyone to a "satisfactory" level of consumption, certainly not to a level anywhere near that of an affluent country in Europe or North America. That problem is compounded by the continued rapid growth of the world's population.

—UNDERSTANDING EARTH, CHAPTER 22

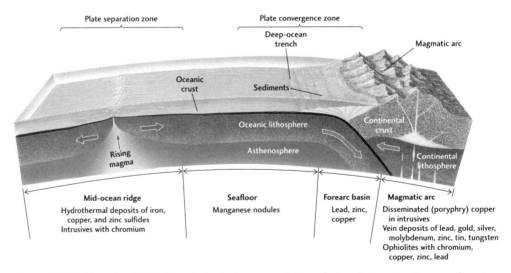

Figure 22.23. The role of plate boundaries in the accumulation of mineral deposits. Ocean sediment and crust are enriched in metals by hydrothermal ore deposition along a mid-ocean ridge. Rising magma in the subduction zone is the source of ores that form the metal-bearing provinces of a magmatic belt such as the Cordillera of North and South America. The melting of subducting sediment and crust may contribute to ore constituents. Mineral-bearing oceanic fragments (ophiolites) accrete to the continent in the collision zone.

Before Lecture

In this chapter, we will consider the following questions. How do these resources form? Where are they found? Who does and will control them? How long will the supplies of these critical nonrenewable resources last, and what will we do when they are exhausted? ... These concerns result from a new and deeper understanding that we cannot continue to draw wealth from the Earth indefinitely without thinking about the consequences for our habitat and for the generations to come.

—UNDERSTANDING EARTH, CHAPTER 22

The quotations suggest that Chapter 22 is a little different from previous chapters in that it seeks to address some very important social concerns for all human beings on the planet. That should make it interesting. But it also poses a learning challenge. It will be important to be aware of your own conceptions and preconceptions as you read this chapter. Pay careful attention when one of your preconceptions is challenged. In other words, try to be open to new ideas and arguments if you encounter them.

Chapter Preview

- **What is the origin of oil and natural gas?**
  Brief answer: Oil and natural gas form from organic matter deposited in marine sediments. The organic matter is pressure-cooked as the sediments are buried and the organic materials are transformed into liquid and gaseous hydrocarbons. Figure 22.5 illustrates how the organic fluids migrate and accumulate in geologic traps.
- **What is the origin of coal and how big a resource is it?**
  Brief answer: Coal is formed by the compaction and mild metamorphism of buried terrestrial wetland vegetation. Coal has supplied an increasing proportion of U.S. energy needs since 1975. Refer to Figures 22.8 and 22.9.
- **What are some of the environmental concerns connected with the use of fossil fuels?**
  Brief answer: Coal mine reclamation, pollution, oil spills, acid rain, and carbon dioxide emissions are major environmental concerns that need to be addressed. Refer to Figure 22.4.
- **What are other alternative sources of energy?**
  Brief answer: Nuclear power can be a major energy source but only if its costs do not keep escalating and the public can be assured of its safety. Hydroelectric and geothermal energy, solar power, wind power, and biomass power are also sources of energy.
- **What is an economical mineral deposit?**
  Brief answer: Mineral deposits become ore deposits when they are rich and valuable enough to mine economically. See Figure 22.1.
- **How do ore deposits of metal-bearing minerals form?**
  Brief answer: Hot water (hydrothermal) associated with volcano, metamorphic, sedimentary, weathering, and igneous processes can enrich metal-bearing minerals to form economical deposits.
- **How are natural resources related to plate tectonics?**
  Brief answer: Many metal ore deposits are formed by magmatic and hydrothermal processes that are closely linked to both modern and ancient plate boundaries. Refer to Figure 22.23.

Vital Information from Other Chapters

The material in Chapter 22 may demand less active review time than previous chapters. The geology is straightforward. Still, the formation of energy and material resources involves a wide variety of geologic processes including magmatic, metamorphic, weathering, sedimentary, and plate tectonics processes. Use the previous chapters as resources for answering questions that may come up while reading Chapter 22. The formation of oil, natural gas, and coal involves sedimentary and metamorphic processes that you learned about in Chapters 8 and 9. Toxic and nuclear waste contamination brings us back to issues related to groundwater and stream transport, discussed in Chapters 13 and 14. The section *The Geology of Mineral Deposits* refers to specific minerals that were introduced in Chapter 3. As always, plate tectonics plays an important role in this chapter. For example, to fully understand Figure 22.23, you may need to return to Chapter 2 to review information on seafloor spreading, forearc basins, and magmatic arcs.

> *Our entire society rests upon and is dependent upon our water,*
> *our land, our forest, and our minerals. How we use these resources*
> *influences our health, security, economy, and well being.*
> —JOHN F. KENNEDY, FEBRUARY 23, 1961

During Lecture

This should be an especially interesting lecture. Earth's natural resources are described in terms of abundance and geological origin. You will not be surprised to learn that plate tectonics plays a role in ore-forming processes. There are many important social issues connected to the material in Chapter 22. Your instructor may make use of a discussion or debate to address some of these issues. Summarize the important social issues and arguments in your notes. Pay particular attention to capturing arguments that contradict your own ideas. Circle such arguments and return to them later for study and consideration. True learning often involves changing our conceptions (and particularly our misconceptions).

After Lecture

The perfect time to review your notes is right after lecture. The following checklist contains both general review tips and specific suggestions for this chapter.

Check your notes: Have you...

- ☐ captured arguments about how we should use our nonrenewable energy and materials source that came up in the lecture or during classroom discussions or debate?
- ☐ created a brief "big picture" overview of this lecture and chapter? Suggestion: See the authors' statement about this chapter on the first page of the text chapter (quoted on the first page of this Study Guide chapter).

Intensive Study Session

Set priorities for studying this chapter. As usual, we recommend that you give high priority to activities that involve answering questions. But this is a unique chapter because it deals with some of the most significant issues and problems that human beings must confront and solve in the next 50 to 100 years. We recommend that you make learning about the social issues of energy and resources your highest study priority in this chapter. Consider using the following strategy.

- **First, get an overview of resource use as a significant social issue.** This is a chapter with unusually powerful implications for our future as human beings. To gain an overview of those implications, read the section titled *Energy Resources and Energy Policy* and examine Figures 22.1, 22.2, 22.3, 22.6. 22.13, and 22.15. Then read the two *Earth Policy* boxes. *Earth Policy* box 22.1 deals with toxic and nuclear waste contamination and *Earth Policy* box 22.2 deals with arguments related to our use of federal lands that make up about one quarter of our country. Think about the implications of these figures. It is time well spent.
- **Preview the key figures.** Next, work on the geology of Chapter 22 by reviewing the key figures that explain the formation of oil (Figure 22.5), coal (Figure 22.8), hydrothermal deposits (Figure 22.16), and how ore deposits relate to plate tectonics (Figures 22.22, 22.23, and 22.25). You have to understand these figures to answer the **Review Questions.**
- **Then try the Review Questions.** Try answering each of the review questions to check your understanding of the lecture. Check your answers as you go, but do try to answer each question before you look at the answer. Notice the test-taking tips that are interspersed with these questions. They are designed to help you do better on your next exam.
- Sometime before your midterm answer all the exercises at the end of the chapter. These are short answer and won't take long if you know the material. Notice that helpful animations are provided on the Web site for some of the chapter exercises.
- **Web Site Study Resources**
  http://www.whfreeman.com/understandingearth
  Complete the **Concept Self-Checker** and **Web Review Questions.** Pay particular attention to the explanations for the answers. Complete the following **Online Review Exercises:** *Identify the Parts of the Fossil Fuel Cycle, Where do Mineral Deposits Occur?, Understanding our Mineral Resources,* and *Understanding Mineral Consumption and Sustainability*. The **Geology in Practice** exercises illustrate how modern seismic profiles of subsurface oil traps beneath the infamous Teapot Dome in Wyoming are used to find oil.

Exam Prep

Materials in this section are most useful during your preparation for exams.

Final Exam Prep

Each semester in one short week you take an exam in each and every course. Most of those exams are comprehensive finals that cover the entire semester. Dealing with finals week successfully can be a challenge. Here are some tips that will ensure that you do your best work during finals week.

Tips for Surviving Finals Week

- Be organized and systematic. Use the **Final Exam Prep Worksheet (Appendix B)** to help you get organized for finals. Use the **Eight-Day Study Plan (Appendix A)** for every course where the final exam will be an important factor in determining your grade.
- Stick to priorities. Say no to distractions.
- Build in moments of relaxation: Take regular short breaks, exercise, and be sure to get enough sleep.
- Be confident. By now you have built up a good set of study habits. You are a competent learner.

The following **Chapter Summary** and **Review Questions** should simplify your chapter review. Read the **Chapter Summary** to begin your session. It provides a helpful overview that should refresh your memory.

Next, work on the **Review Questions.** To determine how you stand on mastery of this chapter, complete the exercises and questions just as you would a midterm. After you answer the questions, score them. Finally, review any question you missed. Identify and correct the misconception(s) that resulted in your answering the question incorrectly.

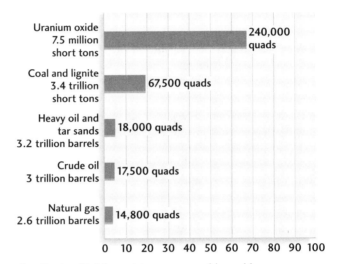

Figure 22.2. A rough estimate of total remaining nonrenewable world energy resources amounts to about 360,000 quads. Amounts are given in conventional units of weight (short tons), volume (barrels), and energy content (quads). A short ton is 2000 pounds, or 907.20 kg; a barrel of oil is 42 gallons. Coal and lignite resources, for example, amount to 3.4 trillion short tons, equivalent to 67,500 quads, or 19 percent of total energy resources.

Chapter Summary

What is the origin of oil and natural gas?

- Oil and natural gas form from organic matter deposited in marine sediments. The organic materials are buried as the sedimentary layers grow in thickness. Heat, pressure, and bacterial action transform the organic matter into fluid hydrocarbons. The fluid hydrocarbons tend to migrate out of the source rock and accumulate in geologic traps that confine the fluids within impermeable barriers. Refer to Figures 22.2, 22.4, and 22.5.
- Petroleum resources will be significantly depleted within about a century.

What is the origin of coal and how big a resource is it?

- Coal is formed by the compaction and mild metamorphism of buried wetland vegetation.

    The process by which coal forms begins with the deposition of plant matter. Protected from complete decay and oxidation in a wetland environment, the deposit is buried and compressed into peat. Subjected to further burial, peat undergoes mild metamorphism, which transforms it successively into lignite, subbituminous and bituminous (soft) coal, and anthracite (hard) coal as the deposit becomes more deeply buried, temperature rises, and structural deformation may occur.

- Domestic coal resources in the United States would last for a few hundred years at current rates of use(about a billion tons per year.

What are some of the environmental concerns connected with the use of fossil fuels?

- Environmental concerns associated with the use of fossil fuels include mine reclamation, pollution, acid rain, oil spills, and carbon dioxide emissions.

What are other alternative sources of energy?

- Alternative energy resources include nuclear, geothermal, hydroelectric, wind, biomass, and solar. Like fossil fuels, there are significant economic, technological, environmental, and political concerns associated with alternative energy resources.

What is an economical mineral deposit?

- Mineral deposits are considered ore deposits when they are rich and valuable enough to be mined economically.
- Hydrothermal, metamorphic, chemical and mechanical weathering, and sedimentary processes can enrich metal-bearing minerals to form economical deposits.
- Important nonmetallic mineral deposits include limestone for cement, quartz sand for glass and fiber optics, gravel for concrete, clays for ceramics, evaporites like gypsum for plaster and wallboard, plus salts and fertilizers.

How do ore deposits of metal-bearing minerals form and how are such resources related to plate tectonics?

- Many metal ore deposits are formed by magmatic and hydrothermal processes, which are closely linked to both modern and ancient plate boundaries. Knowledge of how mineral deposits form and their association with plate boundaries has greatly facilitated the discovery of new deposits.

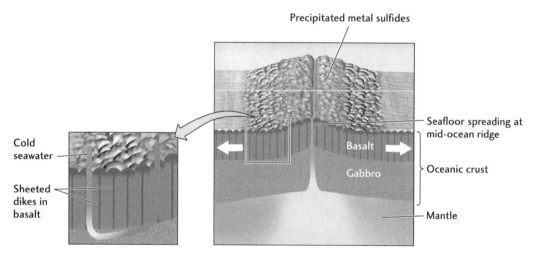

Figure 22.22. Enormous quantities of sulfide ores are found at mid-ocean spreading centers.

Review Questions

Answers and explanations are provided at the end of the Study Guide.

1. What is the sequence from low to high grade for the transformation of plant matter into coal?
    A. Plants, peat, lignite
    B. Peat, lignite, bituminous, anthracite
    C. Bituminous, anthracite, peat, lignite
    D. Anthracite, bituminous, lignite, peat

Test-Taking Tip: Use what you know to guess what you don't know

When confronted with an exam item like number 1, you often don't have to remember the entire sequence to get the item correct. Suppose, for example that you remember for sure that anthracite is the final high-grade product in the series. In that case you can check B with confidence even if you can't remember the rest of the sequence.

2. Which one of the following are energy sources is NOT fossil fuel?
   A. Natural gas
   B. Coal
   C. Uranium
   D. Oil

3. Which of the following is NOT a consequence of fossil fuel consumption?
   A. Mine reclamation
   B. Ozone depletion in our atmosphere
   C. Disposal of residual ash from the burning of coal
   D. Acid rain

4. Oil and natural gas are mostly found in sedimentary rocks deposited in
   A. the deep ocean.
   B. river deltas and on the continental shelf.
   C. wetlands.
   D. large lakes.

5. All the following EXCEPT _____ are effective oil traps.
   A. faults
   B. anticlines
   C. salt domes
   D. horizontal sedimentary and volcanic beds
   Hint: Refer to Figure 22.5.

6. When was the first oil well drilled in America?
   A. 1859
   B. 1880
   C. 1901
   D. 1940

7. The most important source of U.S. energy is
   A. coal.
   B. nuclear power.
   C. oil.
   D. hydroelectric power.
   Hint: Refer to Figure 22.3.

8. How many U.S. gallons are contained in one barrel of oil?
   A. 16 gallons.
   B. 25 gallons.
   C. 42 gallons.
   D. 55 gallons.
   Hint: Refer to Figure 22.2.

9. The United States ranks _____ in oil reserves.
   A. first
   B. second
   C. eighth
   D. tenth
   Hint: Refer to the text section *The World Distribution of Oil and Natural Gas.*

10. Which country has the greatest coal reserves?
    A. China
    B. Great Britain
    C. United States
    D. Former Soviet Union

11. Important factors contributing to the formation of coal from vegetation are
    A. heat and oxidation.
    B. compaction by burial and heat.
    C. biological activity and dissolution.
    D. hydrothermal alteration and metamorphism.

12. From what process is coal derived?
    A. Decay of marine plants and animal matter
    B. Burial, compression, and heating of plant matter deposited in wetlands
    C. Deposition and metamorphism of marine limestones
    D. Transport of organic matter by rivers to their delta
    Hint: Refer to Figure 22.8.

13. Acid rain forms when _____ from the combustion of coal and petroleum combine with rainwater.
    A. hydrogen gases
    B. sulfur dioxide gases
    C. oxygen
    D. nitrogen gases

14. Deposits of gold, diamonds, and chromite found in river gravels and beach sands are classified as
    A. placers deposits.
    B. hydrothermal deposits.
    C. pegmatites.
    D. kimberlites.

15. Metallic ores commonly
    A. form only at converging plate boundaries.
    B. form only at diverging plate boundaries.
    C. form at both converging and diverging plate boundaries.
    D. do not have a strong association with active plate boundaries.

16. The exploration division of World Amalgamated Metals (WAM) has hired you to find its next big copper deposit in the western United States. Where will you target your exploration?
    A. At the base of Triassic conglomerates and sandstones
    B. Within and surrounding Cenozoic igneous intrusive rocks
    C. Deep within Permian limestones
    D. Throughout Pleistocene glacial tills
    Hint: Refer to the *Disseminated Deposits* section in the textbook.

17. Kimberlites are ultramafic igneous rocks from which are mined
    A. copper, lead, and zinc.
    B. nickel, copper, and iron.
    C. chromium and platinum.
    D. diamonds.

18. Evaporites are significant geologic deposits because they
    A. contain gold and silver.
    B. are a major source for plaster board and chemicals.
    C. are an alternative energy resource.
    D. are rich in uranium and other fuels for nuclear power.

19. If you were hired by Dr. Greasy's Mystery Oil Company to locate new oil and gas reservoirs in the western United States, where would you focus your exploratory drilling program?
    A. Where Paleozoic limestones had been metamorphosed to marble by intense magmatism
    B. Where impermeable volcanic rocks cover fractured Precambrian gneiss
    C. Where high angle normal faults cut Mesozoic desert sand dune deposits
    D. Where crystalline rocks had been thrust over unmetamorphosed organic-rich sedimentary strata
    Hint: Refer to Figure 22.5.

20. Generally, by what process are ore mineral vein deposits formed?
    A. Ore minerals are precipitated within fractures and joints when hot metal-bearing fluids are quickly cooled.
    B. Ore minerals are concentrated in veins and channel sediments by rivers and streams.
    C. Ore minerals are concentrated within mud cracks and fractures through surface evaporation.
    D. Ore minerals are concentrated in magma chambers through magmatic differentiation and fractional crystallization.
    Hint: Refer to Figure 22.16

21. Predict which well would have the best potential for oil and gas production.

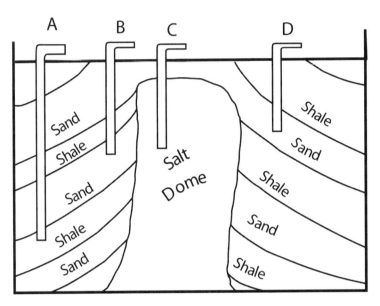

    A. Well A
    B. Well B
    C. Well C
    D. Well D
    Hint: Refer to Figure 22.5.

22. What natural process stores the energy we derive from the burning of fossil fuels?
    A. Photosynthesis
    B. Volcanism
    C. Oxidation
    D. Nucleosynthesis

23. Toxic and radioactive chemicals leaking from the Hanford Superfund site have
    A. luckily remained confined to the Hanford site.
    B. migrated in the aquifer to the Columbia River.
    C. been successfully cleaned up.
    D. reached Seattle.
    Hint: Read *Earth Policy* box 22.1.

24. Commercially important elements are typically concentrated _____ times to produce an economically viable ore deposit.
    A. 2 to 5
    B. 10 to 5,000
    C. a million
    D. a trillion
    Hint: Refer to Table 22.1.

We Americans think we are pretty good!
We want to build a house, we cut down some trees.
We want to build a fire, we dig a little coal.
But when we run out of all these things,
then we will find out just how good we really are.

—WILL ROGERS.

CHAPTER 23

Earth's Environment, Global Change, and Human Impacts

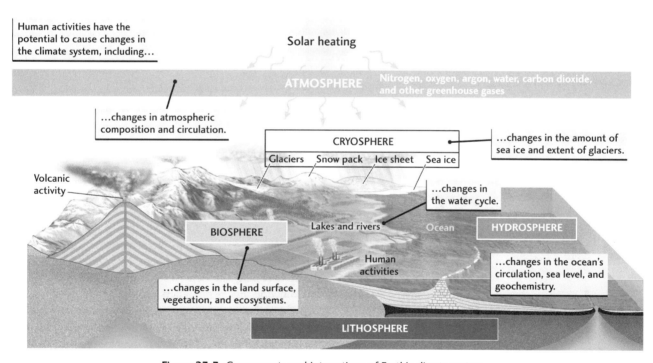

Figure 23.3. Components and interactions of Earth's climate system.

Before Lecture

Before you attend lecture be sure to spend some time previewing the chapter. For an efficient preview use the questions below.

Chapter Preview

- **What is the climate system?**
  Brief answer: The major components of the climate system are the atmosphere, hydrosphere, lithosphere, and biosphere. It is important to understand how these systems interact with each other via *feedback mechanisms*. Refer to Figure 23.3.
- **What is the greenhouse effect?**
  Brief answer: Carbon dioxide and other trace atmospheric gases are transparent to sunlight but absorb heat (infrared radiation), which warms Earth's surface environments. Refer to Figure 23.6.
- **How has Earth's climate changed over time?**
  Brief answer: The largest changes are the 100,000-year glacial cycles. In addition, there are significant short-term climate cycles that average 1,000 and 10,000 years. Refer to Figure 23.7 and note the regularity of these cycles. Note also that we are currently living in an exceptional time: the most prolonged stable warm period in the last 400,000 years.
- **What is the carbon cycle?**
  Brief answer: Geochemical cycles trace the flux of Earth's elements, like carbon, from one reservoir to another. The carbon cycle is particularly important because of its strong link to life processes and climate change. Refer to Figure 23.11.
- **What are important issues in global change today?**
  Brief answer: Climate change, ozone depletion, acid rain, and human population growth are issues of vital concern today.
- **How much global warming will there be in the twenty-first century, and what will be the consequences?**
  Brief answer: The range accepted by most experts is from 1.4 to 5.6 degrees celsius. Consequences include raising sea level, shrinking ice caps (Figures 23.4 and 23.18), effects on many natural systems, and the quality of life of humans. Refer to Table 23.1.

Vital Information from Other Chapters

The carbon cycle (Figure Story 7.6) is essential background material for Chapter 23. Also, review the Milankovitch cycles (Figure 16.27), Figure 16.25, and *Earth Issues* boxes 16.1 and 16.2.

During Lecture

- Keep the big picture in mind as you take notes. Chapter 23 tells the story of the Earth system and how global climate results from interactions between four Earth system components: the atmosphere, hydrosphere, biosphere and lithosphere. Human activities are becoming an increasingly important factor influencing how Earth systems function. Focus on understanding the components, fluxes, and feedbacks within each system.
- Because of the social importance of global climate there may be opportunities for discussion/debate activities. Previewing the chapter will prepare you to take part in these activities.

After Lecture

The perfect time to review your notes is right after lecture. The following checklist contains both general review tips and specific suggestions for this chapter.

Check your notes: Have you . . .

☐ written a summary of what is covered in this lecture? Your summary should say something significant about how human activities change the global environment and the potential for global warming during your lifetime. Suggestion: Write a brief position paper on an issue that concerns you. Ask yourself what earth system information in Chapter 23 is relevant to the issue. Try to develop a position that is grounded in reason and consistent with existing by science.

☐ added important visual material to your notes? Suggestions: Draw overview sketches of important Earth systems such as the climate system (Figure 23.3) and the carbon cycle (Figures 23.11 and Figure Story 7.6). To understand how climate changes over time, sketch a simplified version of Figure 23.7 in a manner that clearly shows the 100,000-year, 10,000-year, and 1,000-year cycles.

Intensive Study Session

Set priorities for studying this chapter. We recommend that you give highest priority to activities that involve answering questions. We recommend the following strategy for learning this chapter.

- **Preview the key figures in Chapter 23.** The most important figures in this chapter are Figure 23.3 (climate system), Figure 23.6 (greenhouse effect), Figure 23.7 (long-term temperature variation of Earth), Figure 23.8 (global warming), and Figure 23.11 (carbon cycle). You have to understand these figures to answer the **Review Questions**.

- **Next, complete Practice Exercises 1 and 2.** You will get the greatest return on your study time by working on these figures because they will help you remember the most important ideas in the chapter.

- **Then try the Review Questions.** Try answering each of the review questions to check your understanding of the lecture. Check your answers as you go, but do try to answer each question before you look at the answer.

- **Sometime before your exam, answer all the exercises at the end of the chapter.** These are short answer and won't take long if you know the material.

- **Web Site Study Resources**
  http://www.whfreeman.com/understandingearth
  Complete the **Concept Self-Checker** and **Web Review Questions**. Pay particular attention to the explanations for the answers. The **Online Review Exercise** *Identify the Parts of the Carbon Cycle* will help you inventory major components of this geochemical cycle.

    Did you know that it takes about 2 pounds of coal to produce the energy for you to copy a megabyte of music off the internet and this releases 4 pounds of carbon dioxide into the atmosphere? Learn all about it in the **Geology in Practice** exercises for Chapter 23.

Exam Prep

Materials in this section are most useful during your preparation for midterm and final examinations. The following **Chapter Summary** and **Practice Exercises and Review Questions** should simplify your chapter review. Read the **Chapter Summary** to begin your session. It provides a helpful overview that should refresh your memory.

Next, work on the **Practice Exercises and Review Questions.** To determine how you stand on mastery of this chapter, complete the exercises and questions just as you would a midterm. After you answer the questions, score them. Finally, review any question you missed. Identify and correct the misconception(s) that resulted in your answering the question incorrectly.

Chapter Summary

What is the climate system?

- Major components of the Earth's climate system are the atmosphere, hydrosphere, lithosphere, and biosphere. Refer to Figure 23.3.
- Earth's surface would be much colder without the presence of greenhouse gases, like water and carbon dioxide, in the atmosphere.
- Ocean current transport play a major role in distributing heat across the Earth because water has a very high capacity for storing heat.
- Topography impacts climate by influencing the flow of our atmosphere.
- Volcanic eruptions affect climate by changing the composition of the atmospheric gases and by adding dust and haze that increase the albedo of the atmosphere.
- Various factors may exert a positive or negative feedback on the climate system. In some cases, feedback mechanisms can act to stabilize Earth's climate and in other cases they may destabilize it by amplifying climate change.

What is the greenhouse effect?

- Carbon dioxide and other trace atmospheric gases act like the glass windows in a greenhouse. They are transparent to sunlight and absorb heat (infrared radiation), which warms Earth's surface environments.

How has Earth's climate changed over time?

- The largest changes are the 100,000 years glacial cycles.
- There are also significant short-term climate cycles that average 1,000 and 10,000 years. Refer to Figure 23.7 and note the regularity of these cycles. Note also that we are currently living in the most prolonged stable warm period in the last 400,000 years.

What are geochemical cycles?

- Geochemical cycles trace the flux of Earth's elements, like carbon, from one reservoir to another.
- Understanding the carbon cycle is important because of its strong link to life processes and climate due to the greenhouse effect.
- The release of carbon dioxide by the burning of fossil fuels is having a significant impact on the flux of carbon from the lithosphere into the atmosphere.

What are important issues in global change today?

- Earth is heating up. The impact of global warming due to increasing levels of greenhouse gases from the burning of fossil fuels is a major concern. Refer to Figures 23.6, 23.8, 23.17, and 23.18, Table 23.1, and *Earth Policy* box 23.2.
- Acid rain is produced mainly from emissions of sulfur-containing gases. Acid rain can cause noticeable damage to forest and lake ecology, fabrics, paints, metals, and building materials. Refer to Figures 23.13 and 23.14.
- A well-defined, large-scale zone of ozone depletion has formed within the stratosphere due to complex interactions with chlorofluorocarbon compounds. Stratospheric ozone shields Earth's surface from damaging ultraviolet radiation.

The Montreal Protocol is an international treaty that appears to have successfully dealt with this environmental disaster. Refer to Figures 23.15 and 23.16.

How much global warming will there be in the twenty-first century, and what will be the consequences?

- The range accepted by most experts is from 1.4 to 5.6 degrees celsius.
- Consequences will include rising sea level and shrinking ice caps at the poles as well as species extinction and shifts in habitat and ecosystems. Refer to Figures 23.4, 23.16, and Table 23.1.

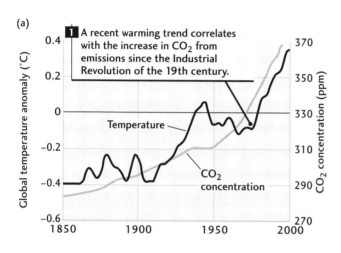

Figure 23.8. (a) From the late nineteenth century to the beginning of the twenty-first, the mean surface temperature has increased by about 0.6 C, a phenomenon called the twentieth-century warming.

Practice Exercises and Review Questions

Answers and explanations are provided at the end of the *Study Guide*.

Exercise 1: Conceptual Map/Flowchart of a Climate Factor

Construct a conceptual map/flowchart characterizing *one* other factor besides clouds that can affect Earth's climate. Also provide a brief written explanation.

Follow these guidelines.

- Be sure you understand and clearly explain how the factor causes climate to change.
- Include in your flowchart possible positive and negative feedback systems. For example, the cooling affect of low clouds, like storm clouds, will reduce evaporation rate and may have short-term negative feedback on additional cloud formation.

Example: Conceptual Map/Flowchart showing how clouds impact climate.

Climate Change Factor: Clouds

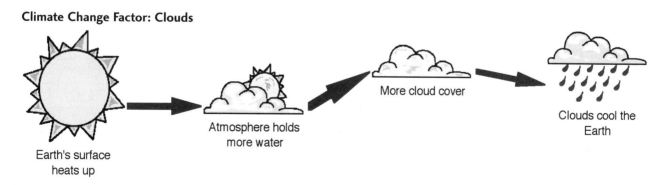

Explanation of Flowchart

In a simple model, as air temperature increases, more moisture is evaporated and held in the atmosphere. This is likely to lead to more cloud cover. Clouds increase the albedo of the atmosphere and may cool the Earth's surface by reducing the amount of sunlight reaching the surface. Increasing cloud cover potentially would have negative feedback on global warming. Clouds may also have a warming effect on the Earth's surface by reducing the loss of heat to space during the night.

Note: How clouds affect climate is still debated. High clouds have been shown to cause warming and low clouds cooling at the Earth's surface. There is growing evidence that clouds have an overall net cooling effect but perhaps not a great one.

Conceptual Map/Flowchart showing how _____ impacts climate.

Climate Change Factor: _____

Explanation of Flowchart

_____
_____
_____
_____
_____
_____
_____

Exercise 2: Release of Carbon Dioxide from Burning of Fossil Fuels

A. Explain why the release of carbon dioxide from the burning of fossil fuels might lead to increased cloud cover.

_____
_____
_____
_____
_____
_____
_____

B. Would an increase in cloud cover represent a positive or negative feedback? Explain. For example, would an increase in cloud cover enhance or reduce the effect of increasing

Earth's Environment, Global Change, and Human Impacts   235

carbon dioxide in the atmosphere? See pages 547–549 in the textbook for examples of feedback mechanisms that can help to balance or stabilize the climate system.

_____
_____
_____
_____
_____
_____
_____

Exercise 3: Flow of Carbon Through Earth's Systems and Reservoirs

Carbon dioxide is an important greenhouse gas that can greatly impact Earth's surface temperatures. Because the Earth is a closed system, the carbon cycle and distribution of carbon in the various reservoirs is a very important component in any model for how climate changes.

Fill out the followng table to summarize the flow of carbon through Earth's systems and reservoirs. Flux is the amount of energy or matter flowing through a given area or reservoir in a given time. Refer to Figures 23.6, 23.11, and 23.12 and the accompanying text in Chapter 23.

Carbon fluxes	Brief description of flux	Direction of flux	Climatic impact/ implications
Life processes	Carbon is fixed in living organisms, which ultimately contribute to organic matter in sediments, coal, and oil.	Carbon flows from the atmosphere and oceans into rock—the lithosphere.	
Sedimentation	Calcium and carbonate ions combine to produce calcium carbonate, which can precipitate and collect to form limestone or help cement other rock particles.		**Climate cools.** Carbon dioxide is drawn out of the oceans and atmosphere. The loss of CO_2 from the oceans will result in a reduction of CO_2 in the atmosphere.
Volcanism			
Chemical weathering	CO_2 in rainwater combines with minerals in rock to form calcium carbonate.	Carbon flows from the atmosphere and oceans into the crust.	**Climate cools.** Carbon is being drawn out of surface environments and stored in the crust. Uplift of high plateaus and mountains may enhance this flux.
Metamorphism	Heating, recystallization, and decomposition of rocks during metamorphism can release large amounts of CO_2.	Carbon flows from the rocks (the crust) into the atmosphere and oceans.	**Climate warms.** Increased levels of CO_2 in the atmosphere enhances the greenhouse effect, which acts to trap heat energy and slow down the loss of heat to space.
Human activities: combustion of fossil fuels	The burning of fossil fuels releases large amounts of CO_2 into the atmosphere.	Carbon flows from the lithosphere (coal, oil, and gas) into the oceans and atmosphere.	

Also refer to this Web site for the latest information on global warming:
http://www.ngdc.noaa.gov/paleo/globalwarming/home.html

Review Questions

Answers and explanations are provided at the end of this Study Guide.

1. Which of the following gasses is most abundant in the Earth's present atmosphere?
   A. Nitrogen
   B. Carbon dioxide
   C. Water
   D. Oxygen

2. The potential for chemical weathering is _____ by acid rain.
   A. not affected
   B. decreased
   C. increased
   D. neutralized

3. Why is carbon dioxide considered a greenhouse gas?
   A. It absorbs heat.
   B. It reflects radioactivity.
   C. It absorbs UV light.
   D. It reflects sunlight.

4. Earth's global temperature trend is clearly
   A. upward over the last century.
   B. downward over the last century.
   C. unchanged over the last few decades.
   D. variable but there has been no overall change.

   Hint: Refer to Figure 23.8.

5. The occurrence of acid rain is most influenced by the
   A. release of radioisotopes by nuclear power plants.
   B. burning of high-sulfur coals.
   C. burning of low-sulfur coals.
   D. weathering of feldspars.

Test-Taking Tip

The presence of two alternatives dealing with the same thing (in this case "sulfur coals") is a hint that one of them is probably the correct answer.

6. Why do scientists suspect that CFCs are the source of ozone depletion in the stratosphere?
   A. CFCs contain chlorine, which reacts vigorously with ozone, while measurements of ozone in the stratosphere show it decreasing at the same time CFCs are increasing.
   B. CFCs form a mixture with volcanic gases in the lower atmosphere, which rises to the stratosphere and reacts with ozone.
   C. CFCs concentrate UV radiation, splitting apart ozone molecules.
   D. CFCs increase the albedo of the stratosphere, which reduces the solar radiation required for the production of ozone.

7. Ozone is a very reactive gas, so as a pollutant in the lower atmosphere ozone presents a significant health hazard. Why does ozone exist in the stratosphere?
   A. It is constantly leaking from the troposphere, where it is produced, up to the stratosphere.
   B. It is formed from the release of gases out of the micrometeoric dusts that bombard the upper atmosphere.
   C. It is formed continuously in the stratosphere by solar radiation and cannot mix or react with other gases because of the thin atmosphere at that altitude.
   D. It is constantly produced from the oceans and rises through the troposphere to the stratosphere.

8. It has been suggested that the uplift of the Himalayan Mountains and the Tibetan Plateau could have contributed to or even caused a global cooling. The link between the Himalayan Mountains and climatic cooling is probably related to
   A. the collision of India with Asia triggering volcanism and increasing the CO_2 concentration in the atmosphere.
   B. the uplift intensifying the monsoon and associated physical and chemical weathering, which resulted in a drawdown of carbon dioxide from the atmosphere.
   C. the fact that high mountains generate more clouds, and their albedo (reflectivity) cools the Earth's surface.
   D. El Niño and the North Atlantic deep-water current.

9. Which of the following is not associated with El Niño events?
   A. Trade winds slackening or reversing direction
   B. Volcanic activity
   C. Change in ocean circulation patterns
   D. Worldwide anomalous weather patterns
   Hint: Refer to *Earth Issues* box 23.1.

10. As the oceans become warmer, _____ CO_2 is released from the oceans into the atmosphere, resulting in a _____ feedback.
    A. more/positive
    B. less/negative
    C. more/negative
    D. less/positive
    Hint: Is CO_2 more or less soluble in warm water? Refer to Chapter 17.

11. The increase of the average temperature on Earth is linked to burning fossil fuels because the
    A. burning process consumes oxygen.
    B. burning process consumes CO_2.
    C. burning process generates CO_2.
    D. smoke given off by burning insulates the Earth.

12. The surface temperatures on Venus, Earth, and to a lesser extent Mars are all well above what can be explained by their distance from the Sun. What other factor significantly contributes to elevated surface temperatures for these inner planets?
    A. Presence of greenhouse gases, like carbon dioxide
    B. Interior heat
    C. Dust from windstorms and volcanoes, which acts to trap heat
    D. Presence of argon and nitrogen in the atmosphere

13. If you were looking for patterns in the fossil record to support the role of global climate change as a factor in large-scale extinctions, what might you expect to see?
    A. Latitudinal patterns in extinctions could exist, where tropical organisms are decimated, while temperate organisms experience only modest losses.
    B. Mass extinctions of land animals would be found, with little impact on marine animals.
    C. Mass extinctions of marine organisms would by seen, with little impact on land organisms.
    D. No pattern would exist; mass extinctions rule out climate change because climate change is too gradual over geologic time.

14. At present the greatest flux of carbon dioxide occurs between our atmosphere and
    A. oceans.
    B. volcanoes.
    C. living organisms.
    D. humans.
    Hint: Refer to Figure 23.11.

15. Which of the following contributes the most carbon to the atmosphere?
    A. Human deforestation and agriculture
    B. Plant uptake of carbon.
    C. Ocean air gas exchange
    D. Humans burning fossil fuel
    Hint: Refer to Figure 23.12.

The balance of evidence suggests a discernible human influence on global climate.
—UN CLIMATE COMMITTEE, IPCC, 1995, P. 4.

It is virtually impossible to change one thing in a complex system without affecting other parts of the system, often in as yet unpredictable ways.
—ANONYMOUS

APPENDIX A

Eight-Day Study Plan*
(Make additional copies to use for every exam you take.)

Here is a guide you can use to prepare for your exam. Everyone develops their own approach to preparing for exams; feel free to adapt these ideas to your particular needs and situation.

The basic idea is to conduct your preparation in a systematic fashion with focus on the most important material. Our plan accomplishes this by dividing the material equally and suggesting how to incorporate the Exam Prep materials provided in this study guide for each text chapter. You begin the plan eight days prior to the exam.

8 Days Before the Exam: Get Organized!

Step 1: Clarify the task. Determine what type of exam you will take by briefly answering the following questions:

1. This exam will cover (list each chapter to be covered):

   _____
   _____
   _____

2. Material and kinds of skills to be particularly emphasized (list chapters/ideas/skills your instructor said would be particularly important):

   _____
   _____
   _____

3. Question format will be (check one that applies):
   ☐ Multiple Choice
   ☐ True False
   ☐ Essay
   ☐ Thought problems
   ☐ Other (specify)_____

---
*Adapted with permission from the University Learning Center, University of Arizona.

4. Review session is scheduled for (enter date here and be sure to attend):_____

Step 2: Divide the material you must review into four equal Parts: A, B, C, and D.

7 Days Before the Exam: Begin your review.   Review all material in Part A.

Do the following for each chapter you review.

1. **Chapter Summary.** To get yourself started, read the chapter summary (Exam Prep section of this guide) for the chapter you want to review.
2. **Practice Exam Questions.** Answer Practice Exam Questions (Exam Prep section of this guide) to see where you are with the material. Force yourself to answer all questions for the chapter without referring to the answer key. Correct only after you have tried all items. Be sure to review carefully any items you missed. Correct the misconception that resulted in the error.
3. **Class Notes.** Review your class notes and annotations you made in the text margin by asking yourself questions.
4. **Focus on visual materials and key figures.** This may also be a good time to redo some of the Practice Exercises in this study guide. Many Practice Exercises are designed to help you master the visual concepts of geology; review visual material in your notes. Test yourself by seeing if you can reconstruct key figures from memory.
5. **Self-Test.** Spend as high a proportion of your study time as possible quizzing yourself.

6 Days Before the Exam: Review Part B.   Repeat instructions for Day 7, this time reviewing Part B. *If you have problems with the material, see your instructor at the next open office hour.*

5 Days Before the Exam: Part C.   Repeat instructions for Day 7, this time reviewing Part C. *If you have problems with the material, see your instructor at the next open office hour.*

4 Days Before the Exam: Part D.   Repeat instructions for Day 7, this time reviewing Part D. *If you have problems with the material, see your instructor at the next open office hour.*

3 Days Before the Exam: All.   Review all parts—A, B, C, and D—fully. Prioritize your time. Focus on important material that will be covered. Work hardest where you are least sure of your self. *If you have problems with the material see your instructor at the next open office hour.*

2 Days Before the Exam: All.   Review all parts—A, B, C, and D—fully. Prioritize your time. *If you have problems with the material see your instructor at the next open office hour.*

Night Before the Exam: Be sure you get the amount of sleep you need to be alert and perform at your best. You don't need to cram. Just stay focused.

Zero Hour: You have prepared well. Allow yourself to be confident. Stay focused and confident during the exam. Use your best test taking strategies.

APPENDIX B

Final Exam Prep*

Worksheet

(To Be Completed 3 Weeks Prior to Exams)

1. **Course Sheets:** In your notebook set up four or five separate sheets of paper—one sheet for each course you are taking. At the top of each sheet list the course and the grade you presently have (be realistic, not hopeful).

2. **Date:** List the date of the final under each course name.

3. **Comprehensive Finals:** Mark with a "C" each course with a comprehensive final.

4. **Exam Format:** Identify the format of the exam (multiple choice, essay, etc.) under the date of the final for each course.

5. **Task:** Under this, identify the levels of thinking expected. **Hint:** Previous midterms are your ultimate resource on this question. List all kinds of questions. Estimate what percent of total points will be devoted to each kind of thinking.

    - ✓ Application to real-world situations
    - ✓ Problem solving
    - ✓ Critical thinking
    - ✓ Understanding principles
    - ✓ Memory of basic facts

6. **Rank Finals in Importance:** In the upper right-hand corner of each sheet, rank in order the most critical and important final to the least important final—the final that will make the least difference in your grade. (Be aware of how much impact your final exam has on your overall class grade.)

---

*Adapted with permission from the University Learning Center, University of Arizona.

7. **List What the Test Will Cover:** For each course on each sheet, list everything the test will cover; remember which exams are comprehensive.
   - ☐ Handouts?
   - ☐ Chapters? (Which ones?)
   - ☐ Lectures?
   - ☐ Discussions?
   - ☐ Other?

   Check your syllabi to be sure you have not left our any important material.

8. Draw a line beneath this list. Then, list what you still have left to do for that particular course.
   - ☐ Which chapters do you still have to preview?
   - ☐ Which lecture notes do you need to review and update?
   - ☐ Which Practice Exercises and Review Questions do you need to complete?
   - ☐ Which labs do you still have to finish?
   - ☐ What papers do you still have to write?

9. Draw another line. Now list the test preparation strategies you will use to study for the exam—study groups or study patterns, self-questioning using the annotations, mapping, charting, questions and answers, concept cards, going over old tests and quizzes, and making up your own problems.

10. Now fill in the calendar by identifying exams, finals, and due dates for papers. Each day you need to do something from No. 8, but you will also need to study and review for the finals at least 2 hours a day. Be sure to use all of your available times—weekends, waiting time, etc.

Work Toward These Goals:

- ☐ Finish all work under No. 8 (Chapter Previews, Practice Exercises, Review Questions, etc.) one week prior to your first final. Review all lecture notes by asking yourself the questions out loud or by having someone quiz you five days prior to your first final (allow 2–3 hours).
- ☐ Divide the work that remains so that you do an Eight-Day Study Plan for each course that you assigned a high priority in No. 6.
- ☐ Remove the distractions from your life. This is not the week to be captured by TV or other addictions. Stick to your priorities. Tell friends and family that you need to focus all your energy on your finals until they are over.
- ☐ Avoid burnout. Build time into your schedule for adequate sleep, relaxation, and exercise.

Answers to Practice Exercises and Review Questions

Chapter 1

Answers to Practice Exercises

Exercise 1: The Evolving Early Earth

```
┌─────────────────┐
│ The Sun and     │
│ planets form.   │
└─────────────────┘
         │
         ▼
┌──────────────────────────────────────────┐
│ Earth's interior heats up due to:        │
│ • decay of radioactive elements          │
│ • Moon-forming impact and other collisions│
└──────────────────────────────────────────┘
         │
         ▼
┌──────────────────────────────────────────┐
│ Consequences of the Earth heating up include:│
│ • formation of an iron core              │
│ • differentiation of the mantle and crust│
│ • release of gases from the interior to form the│
│   oceans and atmosphere.                 │
└──────────────────────────────────────────┘
```

Answers to Review Questions

1. A. A hypothesis is a tentative explanation which can help focus attention on plausible features and relationships of a working model. If a hypothesis is eventually confirmed by a large body of data, it may be elevated to a theory. Theories are abandoned when subsequent investigations show them to be false. Confidence grows in those theories that withstand repeated tests and are able to predict the results of new experiments.

2. D. Gravitational attraction caused the dust and condensing material to collide and clump together (sticky collisions). Refer to Figure 1.3.

3. B. The heat generated by the decay of radioactive elements continues to heat the Earth today.

4. C. Volcanic degassing accounts for all major gases in the Earth's atmosphere except oxygen. Oxygen is a by-product of photosynthesis. Refer to Figure 1.8.

5. B. Figures 1.5 and 1.12 provide timelines for the early history of the Earth.

6. B. The Earth's crust, mantle, and core are thought to have formed when much of the Earth melted. Melting allowed materials within the Earth to segregate (differentiate) according to their density—heavier matter sinks towards the center. Refer to Figure 1.6.

7. D. Refer to Figure 1.11.

8. A. The innerterrestrial or rocky planets are distinctly more compact—denser—than the outer gaseous planets. Saturn has a density less than liquid water. If an ocean were big enough, Saturn would float.

9. D. Refer to Figure 1.5.

10. C. Refer to Figure 1.1 and page 4 in the text. A hypothesis is a tentative explanation which can help focus attention on plausible features and relationships of a working model. If a hypothesis is eventually confirmed by a large body of data, it may be elevated to a theory.

11. B. Confidence grows for those hypotheses that withstand repeated tests and are able to predict the results of new experiments.

12. A. Refer to pages 4–5 in the text.

13. C. Refer to Table 1.1.

14. A. Refer to Figure 1.11.

15. D. Some atmospheric oxygen molecules diffuse into the upper atmosphere where solar radiation transforms them into ozone. Ozone absorbs UV radiation before it reaches the surface, where it can damage plant and animal cells. Oxygen is not essential to all life. In fact, some life find oxygen toxic. Because oxygen gas is very reactive, its presence probably would have reduced the likelihood for the chemical evolution of life in the earliest history of Earth. Lucky for us, the Earth began as an oxygen-free planet. Oxygen readily reacts with hydrogen to form water, but the amount of free hydrogen in Earth's atmosphere is miniscule; therefore, this reaction is not significant.

Chapter 2

Answers to Practice Exercises

Exercise 1: Characteristics of Active Tectonic Plate Boundaries

Complete the following table below by filling in the blank spaces and boxes.

Characteristics	Divergent See Figures 2.5, 2.6, 2.7, and 2.8.	Convergent See Figures 2.5 and 2.9.			Transform See Figures 2.5 and 2.10.
		Ocean/ Ocean	Ocean/ Continental	Collision	
Examples	Mid-Atlantic Ridge, African Rift Valley, Red Sea, and Gulf of California	Japanese Islands Marinas Trench Aleutian Trench	Peru-Chile Trench Andes Mountains Cascade Range	Himalayas and Tibetan Plateau	San Andreas Fault
Topography	Oceanic ridge, rift valley, ocean basins, ocean floor features offset by transforms, seamounts	Trench, island arc	Trench, volcanic arc, and high mountains	Very high plateau and mountains	Offset of creek beds and other topographic features that cross the fault
Volcanism	Present	Present	Present	Not characteristic	Not characteristic

Exercise 2: Conceptual Flowchart or Diagram Illustrating How Plate Tectonics Works

(6) Some hot mantle melts and erupts at **divergent boundaries** to form new ocean crust/**lithosphere.**

(7) Ocean lithosphere starts out hot and high standing. **Gravity pulls** the newly formed lithosphere off the high spots, typically forming a very broad linear ridge, called a spreading center or oceanic ridge.

(8) Oceanic lithosphere slowly cools and becomes thicker, denser, and heavier. Within about 200 ma it becomes heavy enough to sink. **Gravity pulls** the lithosphere down and/or the younger lithosphere above helps **push** older lithosphere down.

(5) A warm, soft, plastic upper mantle, called the **asthenosphere,** decouples the lithosphere from the mantle below and allows the plates to slide under the force of gravity.

(4) Hotter portions rise to the surface.

(3) Cooler, more dense matter sinks.

(9) Subducting lithosphere at **convergent plate boundaries** may eventually sink to the core/mantle boundary completing a convection cycle.

(2) Internal heat results in density differences. Hotter portions of the mantle are less dense, more buoyant, and therefore rise. Cooler portions are more dense and sink.

(1) Convection distributes Earth's internal heat by the physical transfer of hotter material upward while cooler material sinks.

Answers to Review Questions

1. A. Volcanism at the oceanic ridges builds the seafloor.
2. A. Refer to Figure Story 2.11 and Figure 2.14.
3. B. Today, the rate of plate motion varies from a few centimeters to 24 centimeters per year.
4. D. The typical plate motion along most transform faults is horizontal slip (shearing). However, along curves in the transform fault, transextension (forming a depression) and transcompression (forming mountains) may be generated.
5. A. Refer to Figure 2.6b.
6. C. The Hawaiian Islands formed over a hot spot in the middle of the Pacific ocean plate. Some hot spots such as in Iceland, are located coincidentally adjacent to a spreading center. Most hot spots on Earth are not directly associated with plate boundaries.
7. A. The Atlantic coast of North America is a passive continental margin, which is not associated with an active plate margin.
8. D. Pangaea began to breakup during the Jurassic Period. Refer to Figure 2.15.
9. B. Mid-ocean ridges or spreading centers are divergent boundaries where the crust is extending (pulling apart) and mafic magmas are intruding upward from the asthenosphere to feed basaltic volcanism that is building new ocean floor.
10. A. Refer to Figure 2.9(c).
11. C. Volcanism is not characteristic of transform plate boundaries. Some volcanism, typically minor amounts, may occur in association with transform faults where transextension is occurring.
12. C. Refer to Figure 2.15.
13. D. See Figure 2.15.
14. A. Refer to Figure Story 2.11.
15. B. Note that this becomes more interesting when you compare the age of the seafloor to the oldest rocks found on the continents, which are 4.0 billion years old. Refer to Figure 2.14 and pages 40–41 on seafloor isochrons.
16. B. Refer to Figure 2.14.
17. A. Refer to Figure 2.9.
18. B. Refer to Figures 2.6 and 2.9.
19. C. Refer to Figure 2.6(c).
20. A. Refer to Figure Story 2.11.
21. B. Refer to page 44.
22. D.

Chapter 3

Answers to Practice Exercises

Exercise 1: Crystal Structures of Some Common Silicate Minerals

A. single chains of silica tetrahedra
   pyroxene (augite)
   two good cleavage planes intersecting at about 90°

B. double chains of silica tetrahedra
amphibole (hornblende)
two good cleavage planes intersecting at about 60° and 120°
C. sheets of silica tetrahedra
mica (muscovite)
excellent cleavage in one direction

Exercise 2: Major Mineral Classes

1. *silicate (double-chain)*
2. oxide of magnesium and aluminum
3. silicate (sheet)
4. calcium carbonate
5. hydrated calcium sulfate
6. silicate (framework)
7. native element (carbon)
8. calcium/magnesium carbonate
9. silicate (framework)
10. oxide of aluminum
11. sodium halide
12. iron oxide
13. *native element (carbon)*
14. iron sulfide
15. silicate (sheet)
16. calcium sulfate
17. silicate (single-chain)
18. iron sulfide
19. oxide of aluminum
20. silicate (framework)
21. native element (silver)
22. silicate (framework)
23. silicate (isolated tetrahedra)
24. silicate (sheet)

Exercise 3: Identifying Minerals Using Their Physical Properties

Mineral A: muscovite
Mineral B: pyrite
Mineral C: malachite
Mineral D: orthoclase
Mineral E: calcite
Mineral F: gypsum

Answers to Review Questions

1. D. Chloride is an anion and easily gains an electron from sodium which loses and electron to become a cation. With one extra electron chlorine has the electron configuration of the noble gas argon. Upon losing an electron to chlorine, sodium has the electron configuration of the noble gas neon.
2. D. Graphite and diamond are both composed of pure carbon but have significantly different crystal structures. Refer to Figure Story 3.11 in the textbook.
3. B. A rock is an aggregate of one or more minerals.
4. C. All minerals are crystalline solids.
5. D. By definition, minerals are inorganic. Are graphite and diamond minerals? Yes. They are both made from pure carbon—a common element in organic material. However, even though they are made of carbon, graphite and diamond are not produced by biological processes. Carbon is a common naturally occurring element. Graphite is typically found in metamorphic rocks and diamond originates in the Earth's mantle.

248 Answers to Practice Exercises and Review Questions

6. B. Glass is an amorphous material that lacks a crystal structure. Native copper, diamond, and water ice all fit the definition of a mineral.

7. B. Silicate minerals are the most common mineral group in the Earth's crust and mantle. The Earth's core is thought to consist mostly of an iron-nickel alloy.

8. B. Refer to Figure Story 3.11 (h).

9. C. Mica, similar to muscovite in Figure Story 3.11 (g), is a sheet silicate.

10. B. Refer to Table 3.1 and Appendix 3.

11. A. Clay minerals have a sheet silicate structure.

12. A. Refer to Figure Story 3.11 (e and f).

13. B. Quartz is a silicate and calcite is a carbonate.

14. B. Cleavage is the tendency for minerals to break along planes of weaker chemical bonds within their crystal structure.

15. A. The physical characteristics of a mineral are determined by its composition, the nature of the chemical bonds, and the crystalline structure. Although graphite and diamond are both pure carbon, their crystal structures and the chemical bonds within the crystal structure are significantly different.

16. B. Chemical bonds typically exhibit a mixture of ionic and covalent characteristics. Bonds with a more covalent character are stronger and the bond length is shorter. Bonds with a more ionic character are weaker and the atoms tend to be farther apart.

17. D. Cations of similar sizes and charges tend to substitute for one another and to form compounds having the same crystal structure but differing chemical composition. Cation substitution is common in silicate minerals.

18. A. Chemical bonds between carbon atoms within diamond are predominantly covalent. Compared to ionic bonds, covalent bond length is shorter and covalent bond strength is higher.

CHAPTER 4

Answers to Practice Exercises

Exercise 1: The Rock Cycle

A. Plate tectonic settings for the generation of magma are
hot spots/mantle plumes, such as Hawaii
divergent boundaries
convergent boundaries

B. Cooling rates of igneous rocks

Types of igneous rocks	Cooling rates	Textures
Extrusive	Fast cooling	Fine-grained
Intrusive	Slow cooling	Coarse-grained

C. Agents for the transport and deposition of sediments on land are:
ice
water
wind

D. Two processes that transform loose sediments into rock are:
compaction (burial)
cementation
E. What are the two main types of sedimentary rocks and what are they made out of?

Sedimentary rock type	What are they made out of?
Clastic	Rock and mineral fragments
Chemical/biochemical	Precipitation of minerals previously dissolved during weathering

F. What are the four major conditions (geologic settings) that result in metamorphic rocks?
Contact metamorphism associated with intrusions of magma
Regional metamorphism associated with plate collisions
Ultra-high-pressure metamorphism deep within the lithosphere
High-pressure, low-temperature metamorphism associated with subduction zones
G. No. The rock is not melted during metamorphism, although a minor amount of melt "sweat" may be generated during high-grade metamorphism. Igneous rocks are formed from the solidification of melts (magmas).

Answers to Review Questions

1. B. Plutonic rocks solidify from melts, called magmas. Refer to Figure 4.2 and Figure Story 4.9.
2. C. Most magmas are generated from the melting of silicate rocks within the Earth's crust and mantle. On rare occasion, a magma composed of carbonates erupts on the Earth's surface. On Io, the moon of Jupiter, sulfur magmas erupt from about 10 active volcanoes. Io's eruptions can shoot fountains of sulfur compounds 360 km high.
3. A. An igneous rock with a coarse texture, where individual mineral grains (crystals) are visible without magnification, forms when the rock crystallized slowly beneath the Earth's surface. Solidification of a magma body may take 10s to 100s of thousands of years within the crust and millions of years more to cool after completely crystallized. To expose a coarse-grained igneous (plutonic) rock at the Earth's surface requires significant uplift and erosion of the rocks that once sat on top.
4. D. Refer to pages 77–78 in the textbook.
5. A. Igneous rocks solidify from a molten mass, called magma. The cooling rate influences the rate of crystallization and, thereby, the size of crystals within the rock.
6. A. Bedding or layering is characteristic of sedimentary rocks. Refer to *Earth Issues* Box 10.1 on the Grand Canyon. Some volcanic (extrusive) igneous rocks, like lava flows and ash deposits, can also occur in layers.
7. D. Refer to Figure 4.5.
8. D. Refer to Figure 4.6.
9. A. Refer to Figure Story 4.9.
10. C. Mountains are typically formed along convergent plate tectonic boundaries where ocean lithosphere subducts or continents collide.
11. A. Refer the section of text entitled: *Earth's Unique Systems and Rock Cycle*.

Chapter 5

Answers to Practice Exercises

Exercise 1: Igneous Rock Textures

Texture term/sketch description

    A. *phaneritic*/coarse-grained, large interlocking crystals.

    B. phaneritic/visible crystals but not as coarse grained as sample A.

    C. porphyritic/mixed cooling history has large and very small crystals.

    D. aphanitic/fine grained—crystals may not be visible without magnification, looking similar to tiny dots in a sketch.

Exercise 2: Distribution of Igneous Rocks within the Earth

Major layer within the Earth	Example of an igneous rock	General compositional group	General chemical composition
Continental crust (For continental crust, there are two appropriate answers.)	*Granite*	Felsic	More Si, Na, K Less Fe, Mg, Ca
	Andesite/diorite	*Intermediate*	Intermediate
Ocean crust	Basalt/gabbro	Mafic	More Fe, Mg, Ca Less Si, Na, K
Mantle	Peridotite	Ultramafic	*Less Si, Na, K* *More Fe, Mg, Ca*

Exercise 3: Predicting the Change in Composition in a Crystallizing Magma

Olivine → calcium-rich plagioclase feldspar and pyroxene with no olivine

    A. Silica content: increased

    B. Iron content: decreased

Explanation: Refer to the bottom of Figure Story 5.5; it illustrates how mafic (iron and magnesium-rich) silicate minerals crystallize first in a cooling magma. As iron gets tied up in the crystallizing solid phase, the remaining liquid becomes progressive enriched in silica.

Calcium-rich plagioclase → sodium-rich plagioclase feldspar and no olivine

    C. Silica content: increased

    D. Iron content: decreased

Explanation: Refer to the bottom of Figure Story 5.5; it illustrates how mafic (iron and magnesium-rich) silicate minerals crystallize first in a cooling magma. As iron gets tied up in the crystallizing solid phase, the remaining liquid becomes progressive enriched in silica.

Exercise 4: Minerals and Magma Solidification

Refer to Figure Stories 3.11 and 5.5, Figure 5.4, and Table 5.2.

    A. LESS. The presence of iron and magnesium in the magma greatly influences the complexity of the silicate structure because they act to "poison" the polymerization of the silica tetrahedra, preventing more complex silica tetrahedra crystalline structures.

Because much of the iron and magnesium is incorporated into the early-formed crystals, as illustrated in Figure Story 5.5, minerals that crystallized later in the history of the solidification of the magma tend to be depleted in iron and magnesium, enriched in silica, and have more complex silicate structures.

B. LAST. Refer to the explanation for A.

C. MORE. Refer to the explanation for A.

Exercise 5: Partial Melting and Magma Composition

A. LOWER. Refer to Figure Story 5.5.

B. DEPLETED. Much of the iron and magnesium in a magma is incorporated into the early formed crystals, as illustrated in Bowen's Reaction Series. Minerals that crystallized later in the history of the solidification of the magma tend to be depleted in iron and magnesium and enriched in silica.

C. ENRICHED. Refer to the explanation for B and Figure Story 5.5.

Exercise 6: Predicting the Composition of Magma Generated in Subduction Zones

A. MORE. As illustrated by the Bowen's Reaction Series, silicate minerals with more silica content have lower melting temperatures. A partial melt will be enriched in silica relative to the igneous rock from which it was generated.

B. LESS. Silicate minerals rich in iron and magnesium have higher melting points and are the last to melt compared to minerals lower in the Bowen's Reaction Series. Therefore, a magma generated from a partial melt will be enriched in silica and depleted in iron and magnesium relative to the bulk composition of the original rock from which the melt was generated.

C. MORE. Table 5.2, Figure Story 5.5, and Figure 5.14 are very helpful. Na and K concentrate in minerals with greater amounts of silica. They are enriched in that they have lower melting temperatures and crystallize late in the cooling history of the magma.

D. LESS. A partial melt will always have more silica and less iron and magnesium than the parent rock from which it is generated. Therefore, it will be less mafic.

Answers to Review Questions

1. A. Many students mistakenly choose B or C as the answer. However, grain size alone is not the only basis for the classification and naming of igneous rocks. Composition (mineral content) is the other criteria for the classification of igneous rocks.

2. D. Rhyolite is an aphanitic volcanic rock and granite is a phaneritic plutonic rock. Their cooling histories and textures are different. However, they have the same general composition (felsic).

3. B. Rhyolite contains the most silica of the rocks listed. Fissure eruptions may be composed of a great variety of lavas; however, they are typically basaltic.

4. A. Gabbro is the intrusive equivalent of basalt, which is extrusive.

5. D. The distinction between intrusive (plutonic) and extrusive (volcanic) rocks is based on grain size; it is determined by the rate of crystallization.

6. A. Pyroxenes, such as augite, are a common mineral in basalts. They are a ferromagnesium mineral with a single-chain silicate crystal structure.

7. A. Refer to Table 5.2, Figure 5.4, and Figure Story 5.5.

8. B. Slow cooling typically produces larger crystals. Rapid cooling produces finer crystals. An igneous rock with a mixture of coarse and fine-grained minerals formed under conditions where the cooling history was mixed. Both volcanic and plutonic rocks exhibit porphyritic texture. A porphyritic volcanic rock erupts from a magma that has begun to crystallize—crystals already exist within the magma. Early-formed crystals are literally carried to the surface by the remaining melt, which cools quickly upon erupting on the Earth's surface. A porphyritic texture in a plutonic rock may be a result of its cooling history or changes in pressure and other conditions that influence crystal growth within the magma chamber.

9. D. Table 5.1 is a good reference for this question.

10. B. If no new magma ejected into the magma chamber, the composition of the magma may change over time because of fractional crystallization and the segregation of the earlier-formed iron-rich minerals by settling from the melt. The remaining magma becomes progressive enriched in silica and depleted in iron and magnesium.

11. C. Calcium-rich plagioclase and olivine are the first minerals to crystallize from a magma with a mafic composition. Refer to Figure 5.4 and Figure Story 5.5.

12. B. A batholith is a very large body of igneous rock. Batholiths are found in the cores of mountain belts.

13. D. Quartz and olivine do not crystallize from the same magma body. If there was enough iron in the original magma, the silica is consumed in the formation of the ferromagnesium minerals and plagioclase feldspars. The magma completely solidifies before pure quartz can crystallize. A parent magma with enough silica to generate quartz will not contain enough iron and magnesium to generate olivine.

14. A. Intrusive (plutonic) rocks are typically coarser-grained due to slower cooling. Extrusive (volcanic) rocks are finer-grained due to quicker cooling rates. Cooling rates are not the only factors that influence the grain size of igneous rocks. Rapid changes in pressure within a magma chamber can induce rapid crystallization, resulting in fine-grained crystals within an intrusive rock. High water content in residual fluids from a solidifying magma can enhance crystal growth and size.

15. C. Refer to the discussion on *Plutons* in the textbook.

16. D. Refer to the discussion on *Plutons* in the textbook.

17. B. Refer to Figures 5.11 and 5.13.

18. B. Figure Story 5.5 illustrates how the residual liquid within a magma chamber becomes progressively enriched in silica and depleted in iron as the crystallization of the earlier-formed minerals use up the available iron.

19. B. Refer to Figure 5.13.

20. D. Subduction at convergent plate margins produces large amounts of andesites. In fact, andesites are named after the Andes Mountains in South America because they are very abundant there. Magmas generated within the upper mantle, such as at divergent plate boundaries and hot spots, are typically mafic. Refer to Figures 5.11, 5.13, and 5.14.

21. B. Refer to Figure 5.12.

22. A. Refer to Figure 5.13.

23. B. This is the most unlikely hypothesis because melting within a mantle plume, such as the hot spot in Hawaii, is thought to be due to decompression of hot ultramafic rock generating basalt.

24. C. To melt a solid in a kitchen you heat it up on the stove. Rocks within the Earth are already relatively hot but most are still solid. Reduction in pressure (decompression) and the addition of water (fluid-induction) cause rocks to melt within the Earth. Refer to Figures 5.13, 5.14, and 5.15.

Chapter 6

Answers to Practice Exercises

Exercise 1: Lava Types—Their Properties, Eruption Styles, Deposits, Landforms, Association with Plate Tectonics, and Hazards.

Lava Types	Basalt (mafic)	Andesite (intermediate)	Rhyolite (felsic)
Properties			
Eruption temperature	1000 to 1200°C	Intermediate	800 to 1000°C
Silica content	Low (≈50%)	Intermediate	High (≈70%)
Gas content	*Low, up to a few percent*	Variable	High (up to ≈15%)
Viscosity	*Low-fluid magma*	Intermediate	High—very viscous
Typical flow velocity	0.7 to 30 m/minute	9 m/day	Less than 9 m/day
Typical flow length	10 to 160 km	8 km	Less than 1.5 km
Typical flow thickness	5 to 15 m	30 m	200 m
Eruption styles	Typically not very explosive	Explosive	Typically very explosive
Deposits	Flood basalt	Lava flow	Obsidian dome
	Fissure flow	Dome	Dome
	Pahoehoe and aa flows	Pyroclastic flow— tuff and welded tuff	Pyroclastic flow— tuff and welded tuff
	Pillow lava		
	Cinder		
Landforms	Shield volcano	Composite volcano	Composite volcano
	Lava plateau	Summit crater	Large caldera
	Cinder cone	Caldera	Summit crate
	Small caldera	Cinder cone	
Association with plate tectonics	Divergent boundaries	Convergent boundaries	Convergent boundaries
	Hot spots		
Hazards	Lava flow	Lava flow	Explosive blast
	Explosive in contact with water	Pyroclastic/ash flow	Hot gases
	Hot gases	Explosive blast	Pyroclastic/ash
		Hot gases	Mudflows (lahars)
		Mudflow	

Exercise 2: Volcanoes at Plate Tectonic Boundaries

Complete this exercise by filling in the blanks adjacent to the list of volcanic areas with the correct match of type of volcano, characteristic magma type, and magmatic (plate tectonic) setting.

Use Figure Story 6.9, Figures 6.10, 6.18, and 6.25, and an atlas as a reference. Chapter 2 and web site links provided at http://www.whfreeman.com/understandingearth will also help you.

Volcano or volcanic area	Type of volcano (shield, composite, caldera)	Magma type (mafic, intermediate, felsic)	Magmatic (plate tectonic) setting divergent, convergent, hot spot
Hawaii	Shield	Mafic	Hotspot/mantle plume
Tonga Islands	Composite	Intermediate and felsic	Convergent/subduction
Columbia Plateau	Flood basalts	Mafic	Hot spot
Santorini (Thera), Greece	Caldera	Intermediate/felsic	Convergent
Mayon, Philippines	Composite	Intermediate/felsic	Convergent
Iceland	Shields/fissures	Mostly mafic	Divergent and hot spot
Yellowstone	Caldera	Intermediate/felsic	Continental hot spot
Krakatoa, Indonesia	Composite/caldera	Intermediate/felsic	Convergent
North Island, New Zealand	Composite	Intermediate/felsic	Convergent
Crater Lake, Oregon	Composite/caldera	Intermediate/felsic	Convergent
Japan	Composite	Intermediate/felsic	Convergent
Aleutian Islands, Alaska	Composite	Intermediate/felsic	Convergent
Mariana Islands	Composite	Intermediate/felsic	Convergent
Kilimanjaro, Africa	Composite	Intermediate/felsic	Continental rift/hotspot
Pinatubo, Philippines	Composite	Intermediate/felsic	Convergent
Katmai, Alaska	Composite and caldera	Intermediate/felsic	Convergent
Mount Rainier, Washington	Composite	Intermediate/felsic	Convergent
Tambora, Indonesia	Composite and caldera	Intermediate/felsic	Convergent
Vesuvius, Italy	Composite	Intermediate/felsic	Convergent

Answers to Review Questions

1. D. Cooling rate determines the size of the mineral grains within lava flows. If minerals already exist in the magma chamber before an eruption occurs, the lava will erupt with larger crystals floating in it and solidify into a volcanic rock with a mixture of grain-sizes, known as a porphyry.

2. B. Table 5.2 provides basic compositional trends for igneous rocks. Rhyolite contains up to about 77% silica. Basalt is mafic with about 50% silica.

3. B. Water makes up 70–95% of the vapor content of a magma, followed by carbon dioxide and sulfur dioxide.

4. B. Refer to Figure 6.8.

5. B. Hawaii is a classic example of a shield volcano. Refer to Figure 6.9 (a).

6. C. Refer to Figure 6.9 (d) in the textbook.

7. B. The higher silica content of rhyolitic lavas results in a much higher viscosity that favors very thick lava flows or domes. Refer to Figure 6.9 (b).

8. D. Refer to Figure 6.9.

9. B. Composite volcanoes are commonly associated with convergent plate boundaries. Partial melting of the subducting slab or materials above the subducting slab typically generates intermediate magmas.

10. B. The Hawaiian Islands are a chain of volcanoes generated as the Pacific Plate moves over a hot spot in the mantle.

11. D. Magmas rich in silica and dissolved gases cause the most explosive volcanic eruptions. Increasing silica content progressively increases the viscosity of the magma. More viscous magmas are more likely to plug the throat of a volcano until pressure builds up high enough to cause an explosive eruption.

12. A. Shield volcanoes are formed dominantly by basaltic lavas that are more fluid and therefore tend to spread out. Composite volcanoes are dominantly constructed from intermediate and felsic lava flows, domes, and pyroclastic flows. Intermediate and felsic lavas are more viscous and form thicker flows than basalt lavas.

13. D. Refer to Figure 6.10.

14. C. Refer to sections on *flood basalts* and *fissure eruptions*.

15. B. Refer to Figures 6.8 and 6.9.

16. D. Refer to Figure 6.19.

17. B. Refer to *Reducing the Risks of Hazardous Volcanoes* in your textbook.

18. A. Basaltic eruptions occur along the ocean ridge system and form the ocean crust.

19. A. Hot spots are surface expressions of magma plumes coming up from the ultramafic mantle. These plumes are basaltic in composition. The composition of hot-spot volcanism may become more felsic if the magma plume rises through continental crust (e.g., Yellowstone). Hot spots such as Hawaii and Yellowstone occur within the middle of crustal plates. Hot spots are also located along divergent plate boundaries, such as in Iceland.

20. B. Andesites (intermediate magmas) erupt commonly at convergent plate boundaries, where oceanic crust is subducting beneath continental crust. One example of where this process is happening today is beneath the Cascade Range of volcanoes that extend from northern California through Washington. Mount St. Helens is within the Cascades.

21. C. Refer to Figure 6.9.

22. B. If a volcano is fed by one large magma chamber without any additional injections of magma from below, then the remaining melt in the magma chamber becomes progressively enriched in silica and depleted in iron and magnesium as the magma solidifies. Younger lava flows are enriched in silica relative to flows that erupted when more of the magma was molten. Having a higher silica content and a lower iron content, younger lava flows will be less mafic and more viscous (less fluid).

23. B. Refer to *Volcanic Deposits* in your textbook.

24. B.

Chapter 7

Answers to Practice Exercises

Exercise 1: Physical and Chemical Weathering

Fill in the blanks in the following flowchart. Refer to Table 7.2 and Figure 7.15.

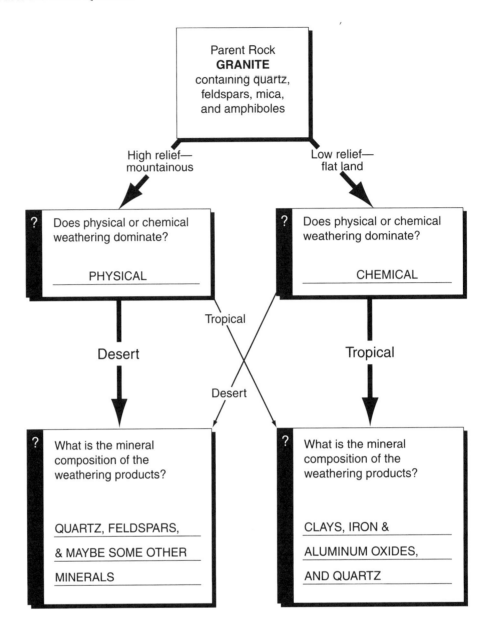

Exercise 2: Soil Types

Fill in the blanks in the following table.

Climate	Soil type	Soil characteristics	Agricultural potential
Desert— warm and dry	*Pedocal*	rich in calcium carbonate in the B-horizon	Higher levels of salts within the soil may compromise the growth of some crops.
Temperate— moderate	Pedalfer	*variable—clay rich*	Pedalfers make good agricultural soils.
Tropical— warm and wet	Laterite	intensely leached soil rich in iron and aluminum oxides	*Lush vegetation is supported by an organic-rich A-horizon. Crops can be grown for only a few years before nutrients are depleted. Source of bauxite, an aluminum ore.*

Exercise 3: Soil Formation in Different Regions

Given the parent rock and location briefly characterize the major soil type (e.g., pedocal, pedalfer, or laterite) and the diagnostic characteristic of each.

A. Soil on a quartz sandstone in semiarid Kansas.

Major soil type: *loose sand/sand dunes—minimum soil development.*

Characteristic(s): *The sandstone will break apart into quartz sand grains and very little else.*

B. Soil on a granite in semiarid Kansas.

Major soil type: Pedalfer.

Characteristic(s): Clay and oxide rich; lacking accumulation of calcium carbonate. Granite contains quartz, feldspars, some mica, and amphiboles. In wetter environments all these minerals but quartz will weather into clay minerals and oxides.

C. Soil on a granite in warm and humid Georgia.

Major soil type: Pedalfer (Laterite)

Characteristic(s): You might be tempted to give laterite for an answer because Georgia has a warm and wet climate. Most of the soils in Georgia are probably between a laterite and a pedalfer in characteristics. They will be clay and iron oxide rich but tend to be nutrient
poor because of the relatively intense leaching due to the high rainfall.

D. Soil on a granite in semiarid southern Arizona.

Major soil type: Pedocal

Characteristic(s): A semiarid to arid climate with low rainfall and high evaporation rates result in slow weathering rates and soil development. Lower rainfall results in less plant growth, less organic matter, and less acidic conditions which further slows weathering rates. Also, soluble salts, such as calcium carbonate, will tend to accumulate in the soil.

Answers to Review Questions

1. D. Oxygen is the principal chemical agent for oxidation reactions. Water is the universal solvent and carbon dioxide combines with water to form carbonic acid. All three play an important role in chemical weathering. Nitrogen gas occurs as a molecule of two atoms of nitrogen. The nitrogen molecule is relatively nonreactive.

2. A. Silica in solution, iron oxides, and clay minerals are all products of chemical weathering. Feldspar is a framework silicate mineral, which crystallized from magma.

3. D. The weathering products of many silicate minerals include clay. However, quartz is pure silicon dioxide and is relatively resistant to weathering. Where chemical weathering is intense, such as in the tropics, even quartz will dissolve.

4. C. Limestone and the calcium carbonate cement of the sandstone are very susceptible to dissolution reactions in wet climates but are more resistant to weathering in dry climates. Shale is a soft sedimentary rock made from compacted clay and typically weathers rapidly. A silica-cemented quartz sandstone might be as resistant to weathering as granite.

5. A. Rusty streaks represent the oxidation of iron-bearing minerals. The other answers are examples of physical weathering.

6. B. Rainfall is naturally acidic due to carbonic acid which forms when carbon dioxide from the atmosphere dissolves in water. Water passing through organic-rich soils dissolves additional carbon dioxide from the decay of organic matter and becomes more acidic.

7. A. Refer to Figure 7.5 in your textbook.

8. D. Table 7.2 shows that calcite, the major mineral in limestones, is very soluble in wet conditions. Also, refer to *Fast Weathering of Carbonates* in your textbook.

9. A. Refer to Figures 7.8 and 7.9.

10. A. Refer to the section in your text titled *The Role of Water in Weathering: Feldspar and Other Silicates* in your textbook.

11. A. The A-horizon is defined as the zone of leaching.

12. C. Chemical weathering is most intense in hot and wet regions.

13. B. The B-horizon is defined as the zone of accumulation.

14. B. Refer to Figure 7.16.

15. C. Pedocals are soils rich in calcium carbonate. Calcium carbonate is relatively soluble in water but it will accumulate in soils where evaporation rates are high in arid and semiarid regions. As the soil dries out, the calcium carbonate is precipitated. Refer to Figure 7.16.

16. B. Soil gases can be rich in carbon dioxide released by the decay of organic matter and life processes. Water percolating down through organic-rich soils becomes more acidic as it dissolves the carbon dioxide in soil gas. Increased acidity of the water enhances chemical weathering reactions.

17. D. Relative to other silicate minerals, olivine is very susceptible to chemical weathering. Its simple (less polymerized) silicate crystal structure of single tetrahedra bonded with cations of iron and/or magnesium is easily attacked by chemical agents. Refer to Table 7.1.

18. B. Chemical weathering is most intense in warm and wet regions.

19. C. Pollution and gases from volcanic eruptions combine with rainfall to enhance its acidity.

Before controls for air pollution were implemented, rainfall in some regions of the United States was as acidic as lemon juice. Acids are powerful chemical agents. Chemical weathering is greatly enhanced by increasing the acidity of water.

20. A. Pedalfers are good agricultural soils. Pedocals can also be farmed. Mineral nutrients in laterite soils are stored in organic-rich A-horizon and can quickly be depleted by farming. Bauxite is aluminum ore and lacks most mineral nutrients necessary for plant growth.

21. B. Climate is the most important factor controlling chemical weathering and soil development.

22. B. Lateritic soils form in tropical regions where chemical weathering is intense.

23. D. As the carbonate cement dissolves, the quartz sand grains will fall away from the rock surface but weathering will have little impact on the quartz grains themselves. Quartz is a very stable mineral on the Earth's surface. Refer to Tables 7.1 and 7.2.

24. C. Soil fertility depends on the availability of the mineral nutrients that are released by chemical weathering from rock forming minerals. When chemical weathering is very slow nutrients remained tied up in the silicate minerals within the rock and are not easily extracted by plant roots. In regions where chemical weathering is intense, most of the mineral nutrients are washed out of the soil. Therefore, fertile soils form where weathering occurs at moderate rates. The parent rock is another important influence on soil fertility. Weak soils developed on a quartz sandstone are likely to be nutrient poor because quartz is pure silicon dioxide and lacks vital mineral nutrients, such as potassium, calcium, iron, and magnesium, for plant growth.

25. D. Granite is composed mostly of feldspars and quartz. Clay minerals are a major product from the chemical weathering of feldspar. Quartz is very resistant to weathering.

26. C. Soils typically take thousands of years to develop.

27. The sandstone obelisk deteriorated so quickly after being moved from Egypt to New York City because the climate in New York is significantly wetter than in Egypt. As naturally acidic rain waters dissolve the calcium carbonate cement, the surface of the sandstone deteriorates. Other factors that may have contributed to the enhanced weathering of the obelisk include acid rain generated by air pollution and frost wedging, as freeze/thaw conditions are common in New York during the winter.

CHAPTER 8

Answers to Practice Exercises

Exercise 1: Common Sedimentary Environments

Environment of deposition	Sediment
Alpine or glacial river channel	Sand and gravel
Dunes in a desert	Fine sand
Flood plain along a broad meander bend	Silt and clay (mud)r
River delta along a marine shoreline	Sand, mud, calcified organisms (seashells)
Continental shelf	Sand and mud
Deep sea adjacent to a continental shelf	Mud
Shoreline beach dunes	Sand, gravel, calcified organisms
Tidal flats	Sand and mud
Organic reef	Calcified organisms

Exercise 2: Grain Sizes for Clastic Sedimentary Rocks

Grain size	Common object	Sediment	Rock type
Coarse	Football to bus	*Boulder gravel*	Conglomerate
↑	Plum or lime	Gravel	*Conglomerate*
	Pea or bean	Gravel	Conglomerate
	Coarse-ground pepper or salt	Sand	*Sandstone*
↓	Fine-ground pepper or salt	Sand	Sandstone
Fine	Talcum powder or baby powder	Mud	Shale

Exercise 3: Clastic and Chemical Sediments and Sedimentary Rocks

Sediment Types	Sedimentary Rocks			
biochemical	arkose	dolostone	limestone	sandstone
clastic	chert	evaporite	peat	shale
chemical	conglomerate	graywacke	phosphorite	siltstone

Statement	Sediment type	Sedimentary rock example
Composed largely of rock fragments	Clastic	Sandstone, graywacke, conglomerate
Precipitated in the environment of deposition	Chemical	Evaporite, phosphorite
Important source of coal	Biochemical	Peat
Often formed by diagenesis	*Chemical*	*Dolostone and phosphorite*
Formed from abundant skeleton fragments of marine or lake organisms, such as coral, seashells, foraminifers	Biochemical	Limestone
Produced by physical weathering	Clastic	Conglomerate, sandstone, siltstone, shale
Produced from rapidly eroding granitic and gneissic terrains in an arid or semiarid climate	Clastic	Arkose sandstone

Answers to Review Questions

1. D. Shale, sandstone, and conglomerate are clastic rocks. Dolomite, chert, limestone, coal, and gypsum are chemical/biochemical.
2. D. See Table 8.3.
3. D. Sand is a term that refers solely to a particular range of grain sizes. It does not imply any specific composition. Quartz and feldspar are common constituents of sand grains because they are common minerals in the Earth's continental crust. Carbonates, such as limestone, dolostone, and fragments of seashells, can also make up sand grains.
4. D. See Figure 8.1.
5. B. There is no known marine life that precipitates dolomite. Dolomite can be precipitated inorganically in seawater, not fresh water.
6. A. Refer to Figure 8.11.
7. A.
8. A.
9. D. Cross-bedding is a feature produced as sediment is deposited by currents of air or water.
10. D. Carbonate sediments are deposited in warm water. Carbonates dissolve in cold water because their solubility is linked to the amount of carbon dioxide dissolved in the water. The solubility of carbon dioxide increases as the temperature of the water decreases.
11. A. Refer to Figure 8.15.
12. B. See Table 7.2.
13. D. Figure 8.15 will help.
14. C. See Table 8.1.
15. D. See Table 8.1. Ultimately, the weathering of most silicate minerals, except quartz, produces clays which become compacted into shale (see Figure 8.11).
16. B. Refer to Figure 8.10.
17. A. Refer to Figure Story 8.16 and *Earth Policy* Box 8.1.
18. D.
19. A.

Chapter 9

Answers to Practice Exercises

Exercise 1: Classification of metamorphic rocks based on texture

Complete the table by filling in the blanks.

Parent rock	Metamorphic rock	Texture (foliated/granoblastic)
Shale	Slate	*Foliated*
Quartz-rich sandstone	Quartzite	Granoblastic
Shale, impure sandstone and many kinds of igneous rocks	*Granulite*	Granoblastic
Granite	Schist & gneiss	Foliated
Limestone	Marble	Granoblastic
Carbonate-rich sedimentary rocks	*Hornfels*	Granoblastic
Basalt (mafic volcanics)	*Amphibolites and greenstones*	Granoblastic
Igneous and metamorphic rocks	*Migmatite*	Foliated

Exercise 2: Comparing Igneous, Sedimentary, and Metamorphic Rocks

Complete the table by filling in the blanks.
   Note that there may be more than one reasonable answer for some blanks.
quartz, muscovite, chlorite, and garnet

Major mineral composition	Texture	Rock type (igneous, sedimentary, metamorphic)	Rock name (granite, sandstone, marble)
Calcium carbonate	*Nonfoliated*	Metamorphic Sedimentary	Marble Limestone
Quartz, K & Na feldspar, mica, and amphibole	*Phaneritic*	Igneous (plutonic)	Granite
Clay	*Fine-grained clastic*	Sedimentary	Mudstone, shale
Pyroxene, calcium feldspar, and olivine	Aphanitic, Porphyritic	Igneous (volcanic)	*Basalt*
Quartz	*Nonfoliated*	Metamorphic	Quartzite
Pebbles and cobbles of a variety of rock types	Clastic	Sedimentary	Cconglomerate
Fragments of seashells and fine mud	Bioclastic, biochemical	*Sedimentary*	Limestone
Quartz, muscovite, chlorite, and garnet	Foliated	Metamorphic	*Schist*

Answers to Review Questions

1. D. If the rock is melted, it is an igneous rock. Some metamorphic rocks get hot enough to "sweat" quartz, such as migamtite.
2. B.
3. B. Refer to section on *Metamorphic Textures*.
4. B. Refer to Figure Story 9.4.
5. A. Refer to Figure Story 9.4.
6. C. Metamorphic rocks characteristically exhibit either foliated or granoblastic textures.
7. D. Although mineralogy may be altered by metamorphism, typically there is little to no change in bulk composition of the rock.
8. D. Even preexisting metamorphic rocks can be metamorphosed.
9. B. Slate → phyllite → schist → gneiss is the correct sequence from fine to progressively coarser-grained metamorphic rocks. All rocks listed in this sequence are foliated.
10. A. Refer to Figure Story 9.7.
11. C. Meteor impacts of the lunar surface are an example of shock metamorphism.
12. D. Refer to Figure Story 9.7.
13. B. Gneiss has the same mineralogical composition as granite. Its distinctive characteristic is foliation. Granite is a common parent rock for gneiss.
14. C. Both slate and gneiss are foliated metamorphic rocks. Being lower in grade, slate is much finer grained.
15. D. Refer to Figure 9.3.
16. A. Refer to the bottom of Figure Story 5.5.
17. B. The intrusion of a hot magma body causes contact metamorphism.
18. B. As illustrated in the diagram at the bottom of Figure 5.5, quartz softens at relatively low temperatures. Oval quartz pebbles could have been stretched into the cigar-shaped features by low-grade metamorphism.
19. C. Refer to Figure Story 9.7.
20. A. Refer to Figure 9.10 and the section in the text on *Continent-Continent Collisions*.

Chapter 10

Answers to Practice Exercises

Exercise 1: Determining the Succession of Geologic Events

A. 12. Even though the black dike does not cut all rock units in the diagram, it does cut the granite and layer 1, which is above all other layered rock units. Therefore, the dike is younger than the granite and layer 1. If the dike was not shown cutting the granite, one would not be able to resolve the relative ages of the dike and the granite. Radiometric dates on the two igneous rock bodies would resolve their ages.

B. 10. Layer 10 is the oldest in the outcrop (diagram). It is at the bottom of a tilted sequence of layered rocks that go from layer 10 (the oldest) to layer 3 (the youngest of the sequence), assuming that this whole sequence has not been completely overturned.

C. Yes. Layer 3 is older than the dike (unit 12). Even though the dike does not cut layer 3, it does cut the unconformity which cuts layer 3 and it also cuts layers 2 and 1, which are above layer 3.

D. Not possible to know. Because the black dike cuts both the granite (unit 11) and layer 1, we can say that layer 1 and the granite are older than the dike. However, the granite may have intruded into the rock layers before or after the deposition of layer 1. There is no way of telling whether the granite is older or younger than layers 4, 3, 2, or 1. A radiometric date on the granite and layer 1, which is a lava flow, would help to resolve this question.

E. 1. Unit 1 is older than unit 12 but younger than unit 11. You do not even have to calculate radiometric ages to answer this question. Just look at the trend in the abundance of the radioactive parent atoms or accumulation of the daughter atoms. The rock with the greatest amount of parent atoms (and least amount of daughter) is the youngest. The rock with the lowest amount of radioactive parent atoms and the largest accumulation of daughter atoms is the oldest. Because the abundance of radioactive parent atoms in unit 1 falls between units 11 and 12, the radiometric age of unit 1 would be some number of years between unit 11 and 12. Unit 11 has the fewest radioactive parent atoms remaining; therefore, it is the oldest.

Exercise 2: Ordering Geologic Events

In the illustration below, a geologic outcrop reveals three layers of sedimentary rocks, one fault and a single igneous dike intrusion.

This is a clear illustration of the principle of superposition and cross-cutting relationships at work. Using the principle of superposition, we see that the limestone must have been deposited before the shale, and the shale before the sandstone. Likewise, these beds must exist before being first cut by the dike and second by the fault.

Youngest	Faulting
	Dike intrudes
	Deposition of sandstone
	Deposition of shale
Oldest	Deposition of limestone

Exercise 3: Marker Events for the Geologic Time Scale

A. Enter each event, listed below, on the line in the appropriate eon box in which the event happened. When possible, order your listing so that the oldest is at the bottom of the list and the youngest is on top.

B. Fill-in the names of eras, periods, and epochs in the correct sequence from oldest at the bottom to youngest on top. Refer to both Figures 1.12 and 10.12 to complete this exercise.

Eon	Era	Period	Epoch
Phanerozoic *Humans evolve* Dinosaur extinction event	Cenozoic	*Quaternary* Tertiary	*Holocene* *Pleistocene* Pliocene Miocene Oligocene Eocene Paleocene
	Mesozoic	Cretaceous *Jurassic* Triassic	
Evolutionary Big Bang	Paleozoic	Permian *Pennsylvanian* Mississippian Devonian Silurian *Ordovician* Cambrian	
Proterozoic Oxygen buildup in atmosphere *First nucleus-bearing cells develop*			
Archeon Earliest evidence of life Major phase of continent formation completed			
Hadean End of heavy bombardment Moon forms *Earth accretion begins*			

Exercise 4: Geologic Time Scale Mnemonic

Construct a mnemonic device for remembering the Geologic Time Scale names. The first letter of each word must match the first letter of the corresponding period or epoch in the proper order. You may use your native language, but be careful not to mix up the words when you do so.

There is no "correct" answer to this question. Any mnemonic device that you will find useful in remembering the different epochs and periods of the Geologic Time Scale is an acceptable answer.

Answers to Review Questions

1. B. The rock layer at the top of an undeformed sequence of rock layers is the youngest in the sequence.

2. D. Cenozoic/Mesozoic/Paleozoic. Refer to Figure 10.12.

3. B. Paleocene/Eocene/Oligocene/Miocene/Pliocene are the epochs of the Tertiary period in the Geologic Time Scale. Remember that the Pleistocene and Holocene are epochs within the Quaternary period. Refer to Figure 10.12.

4. B. Cambrian/Ordovician/Silurian/Devonian/Mississippian/Pennsylvanian/Permian. Refer to Figure 10.12.

5. D. Deposition/deformation/erosion/deposition. Refer to Figure 10.8.

6. A. Rock layer 3 is deposited on top of layer 2 and the 60-million-year-old pluton A. Based on superposition, layer 3 is younger than 60 million years. Layer 3 is also cut by the 34 million-year-old pluton B. Therefore, based on this cross-cutting relationship, layer 3 has to be older than 34 million years.

7. C. Refer to *Radioactive Atoms: The Clocks in Rocks* section of Chapter 10.

8. C. Refer to *Radioactive Atoms: The Clocks in Rocks* section of Chapter 10.

9. D. Refer to Figure 10.12 in your textbook.

10. B. Radiocarbon can be used to radiometrically date carbon-containing materials, such as charcoal, if the sample is younger than about 70,000 years old. Refer to Table 10.1.

11. B. Because radiocarbon has a short half-life (5730 years), the level of radiocarbon decays below the detectable limit in samples older than about 70,000 years. Refer to Table 10.1 in your textbook.

12. C. Given a half-life of one billion years, for every billion years that elapse the number of radioactive atoms decreases by half. Therefore, after one billion years there are 500 left from the original 1000 atoms; after two billion years there are 250 out of the 500 left; and after three billion years there are 125 out of the 250 left.

13. D. To correctly interpret this question it is important to distinguish between observation (data) and inference. The radiocarbon age of 3000 years is data derived from laboratory analysis of the charcoal sample. The result of the analysis represents the approximate time that has elapsed since the organism died (in this case wood) and was no longer exchanging carbon dioxide with the environment. It is reasonable to infer that the charcoal formed when inhabitants at the site burned wood for cooking and heating. It is also reasonable to assume that the wood was not dead for a long time before it was harvested for fuel. If these assumptions are correct, then the radiocarbon date represents the approximate time for the occupation of the site. However, the charcoal could have formed from a forest fire decades or centuries earlier than human occupation. Or the wood harvested for fuel may have been dead and laying on the ground for 100s or even 1000s of years. In arid regions where decay is very slow, dead wood can lie on the ground for 100s to 1000s of years. If "old" wood happened to be used to fuel the fires at this site, then the radiocarbon age could be significantly older than the time of occupation.

14. A. The Geologic Time Scale was constructed over about the last 200 years by geologists using mainly fossils, superposition, and cross-cutting relationships to establish the relative ages for thousands of rock outcrops around the world. In about the last fifty years, the Geologic Time Scale has been calibrated using radiometric methods to date mostly igneous and metamorphic rocks.

15. A. Review *Radioactive Atoms: The Clocks in Rocks* section of Chapter 10.

16. A. Refer to Figure 10.10.

17. A. Loss of daughter atoms, such as lead, generated from the decay of radioactive parent atoms, such as uranium, would result in a date that was younger than the actual time that has elapsed since the rock solidified. One important assumption that is made when interpreting radiometric dates is that the mineral or rock has remained a closed system—no elements have been added to or removed from the mineral except by radioactive decay. If this assumption holds then the radiometric date typically represents the time that has elapsed since the rock solidified from a magma.

18. C. Weathering and alteration of rock samples by metamorphism can cause a redistribution of radioactive parent atoms and the daughter atoms produced by decay of the parent. This redistribution usually causes uncertainty in the radiometric date. Therefore, geologists typically radiometrically date the freshest rock samples that have been the least affected by subsequent geologic events. Sample C in the middle of the lava flow is the least likely to be affected by weathering (sample A) or contact metamorphism (sample B).

19. B. Because the radiometrically dated cobble is included in the conglomerate layer, the layer has to be younger than the cobble—the cobble had to form before it could become a part of the conglomerate. The radiometric date represents a maximum age for the conglomerate, which has to be younger than any included component. If a geologist really wanted to get a better estimate for the age of the conglomerate, they would need to do radiometric dates for as many cobbles of different igneous rocks as possible to find the date of the youngest cobble included in the layer. In this way, one could focus in on the approximate age of the sedimentary layer using the radiometric dates on igneous inclusions and the principle of included fragments.

20. A. Cross-cutting relationships tell us that the dike and lava flow must be the last event to have occurred. Using the principle of superposition we can see the first event must be the deposition of the lower shale, then the limestone-shale-sandstone sequence, so these cannot be the most recent event. Cross-cutting relationships again inform us that the fault and pluton are the next events, though which is more recent than the other is inconclusive, as neither cuts the other. Finally, the dike cuts all the preceding rock units, extruding the lava flow at the surface—the final rock unit. Refer to the sections on *The Stratigraphic Record* in your text for a discussion on these concepts.

21. D. D is the best answer although one might be inclined to answer C based on superposition. Archeologists will commonly use radiocarbon ages to bracket the age of an archeological site if no datable material is found within the site itself. Again, it is important to distinguish between observation (the data) and inferences (the interpretation). The radiocarbon date represents the time that has elapsed since the wood died. Perhaps the wood died during a forest fire and the charcoal was washed away from the burned area and deposited in a layer that covered the fire pit. If this is the case, then the radiocarbon date would be younger than the fire pit. Although charcoal is soft, it is chemically very stable and can remain well preserved in sediments. It is also possible that the charcoal weathered out of some sedimentary layers that are actually much older than the fire pit and end up incorporated in the layer that covers that pit. In this case, the radiocarbon date would be misleading because the charcoal was actually significantly older than the fire pit but ended up in a soil layer above the fire pit because it was recycled from eroding older sediments.

Chapter 11

Answers to Practice Exercises

Exercise 1: Silly Putty®

Compare the properties of Silly Putty® with the behavior of rocks by completing the table below.

Behavior of Silly Putty®	Behavior of rock	Type of force	Geologic structure produced by this style of deformation
Snaps into pieces	Brittle	*Tensional*	Fault or joints
Bends	*Ductile*	Compressional	Folds
Bounces	*Elastic—Rocks do exhibit elastic behavior. More on this when we study earthquakes.*	*Compressional—The ball of putty is compressed by the impact with the floor.*	NOTE: *Earthquakes are attributed to the elastic properties of rocks.*

Answers to Practice Exercises and Review Questions   267

Exercise 2: Geologic Structures

A. reverse fault
B. compressional
C. convergent
D. anticline
E. compressional
F. convergent
G. normal fault
H. tensional
I. divergent
J. strike-slip fault
K. shearing
L. transform
M. syncline
N. compressional
O. convergent

Exercise 3: Anticline vs. Syncline

A. In an anticline the youngest rock layer is on the outside of the fold; in a syncline the youngest rock layer is in the middle of the fold.

B. Refer to Figures 11.16, 11.17, and 11.18.

Exercise 4: Identifying Geologic Structures

A. anticline
B. syncline
C. thrust
D. reverse
E. normal

Answers to Review Questions

1. C. Solids that break are called brittle. Whether a solid behaves brittlely depends on its composition, temperature, confining pressure, and the rate of application of directional forces.
2. A. Refer to Figure 11.4 in your textbook.
3. B. The San Andreas is a right-lateral strike-slip fault. It also represents a transform plate boundary.
4. B. Refer to Figure Story 11.6 and Figures 11.13 and 11.22.
5. B. Joints are fractures along which there is no appreciable movement. Faults are fractures along which there is appreciable offset.
6. D. All other conditions mentioned would favor brittle behavior.
7. A. Basalt is an igneous rock with interlocking fine-grained silicate minerals that have a high melting point. Therefore, basalt tends to behave brittlely. Rocks with high clay or calcium-carbonate content tend to behave ductilely.
8. C. Columnar jointing is fracturing caused by shrinkage during cooling of an igneous rock.
9. D. Normal faults are a result of brittle behavior in response to tension.
10. B. Thrust faults are caused by compressional forces typical of what is generated at convergent boundaries, such as continental collisions.

11. A. Refer to Figure 11.4.
12. B. Refer to Figures 11.17 and 11.18.

CHAPTER 12

Answers to Practice Exercises

Exercise 1: Inventory of the Different Kinds of Mass Wasting

Kind of mass wasting	Composition of slope (consolidated vs. unconsolidated and wet vs. dry)	Characteristics
Rock avalanche	**Large masses of rocky materials**	Speed: running or a speeding auto Slope angle: steep slopes Triggering event(s): earthquakes Notes: Occur in mountainous regions where rock is weakened by weathering, structural deformation, weak bedding, or cleavage planes.
Creep	**Soil**	Speed: **walking** Slope angle: any angle Triggering event(s): none Notes: **Influenced by the kind of soil, climate, steepness of slope, and density of vegetation.**
Earthflows	**Soils and fine-grained rock materials, such as shales and clay-rich rocks**	Speed: **walking or running** Slope angle: any angle Triggering event(s): intense rainfall Notes: Movement is fluid-like.
Debris flow	**Rock fragments supported by a muddy matrix.**	Speed: **running or speeding auto** Slope angle: any angle Triggering event(s): **intense rainfall** Notes: **Contains coarser rock materials compared to earthflows.**
Mudflow	Mostly finer rock materials with some coarser rock debris with large amounts of water.	Speed: **speeding auto** Slope angle: **any angle** Triggering event(s): intense rainfall or catastrophic melting of ice and snow by a volcanic eruption Notes: Contains large amounts of water.
Debris avalanche	Water-saturated soil and rock	Speed: **speeding auto** Slope angle: **steep** Triggering event(s): **earthquakes** Notes: **Occurs in humid, mountainous regions.**
Slump	**Unconsolidated rock material**	Speed: walking or running Slope angle: any slope Triggering event(s): rainfall Notes: **Debris slide moves faster than a slump.**
Solifluction	Surface layers of soil	Speed: walking Slope angle: any angle Triggering event(s): **freeze—thaw** Notes: Occurs only in cold regions when water in the surface layers of the soil alternately freezes and thaws. Water cannot seep into the ground because deeper layers are frozen.

Exercise 2: Water's Role in Mass Wasting

Water enhances and triggers the potential for mass wasting in many ways. Using your textbook as a guide, briefly list five different ways water enhances the potential for mass movements. Below are some sample reasonable answers.

1. Water lubricates, especially if the ground is saturated (i.e., all pore spaces are filled with water), by reducing the internal friction between rock particles. In unconsolidated rock material a small amount of water increases surface tension which actually helps to weakly "glue" the damp, loose material together. Too much water keeps the particles apart and allows them to move freely over one another. Therefore, saturated sand, in which all pore space is occupied by water, runs like a fluid and collapses to a flat pancake shape (refer to Figure Story 12.1).

2. Water (hydrostatic) pressure may become great enough to separate the grains or promote the slippage of beds past one another. Refer to the section in your textbook entitled *Water Content*.

3. Water is a major agent in physical and chemical weathering which promotes mass wasting. Weathering promotes mass wasting by chemically and physically weakening rock.

4. Freezing and thawing are two specific roles water plays in causing solifluction, a kind of mass movement. Refer to the discussion in your textbook on *Unconsolidated Mass Movements*. Undercutting and oversteepening of hillslopes by erosion is another way water enhances mass wasting. Flowing water in rivers and frozen water in glaciers are powerful agents of erosion. Rivers typically erode on the outside of bends in the river. Erosion can undercut and oversteepen the riverbank and adjacent hillslope.

6. Water-saturated rock materials rich in clays or loose sand may be transformed into fluid slurries by a process called liquefaction. Refer to the section in your textbook entitled *Triggers for Mass Movements*.

Exercise 3: Evaluation of Slope Stability

A. Discuss three factors that enhance the potential for mass movement at the home site shown.

- The house is built on a cut-and-fill foundation, where the house is built partly on a cut into the bedrock of the slope and partly on the gravel fill derived from the cut. A cut-and-fill foundation is particularly susceptible to slope failure because it is very loose material that has been bulldozed into place.
- The house and associated possessions add weight to the slope.
- Watering the lawn will enhance the potential for slope failure.
- The presence of a spring indicates that the slope beneath the house is saturated with water. Water adds weight to the slope and acts as a lubricant.
- Traffic on the road below the house adds weight and creates vibrations in the ground, which may compromise slope stability.
- The slope consists of sedimentary rocks that have a dip parallel to the slope. Bedding planes are zones of weakness within these layers; therefore, this rock layer orientation increases the potential for slope failure.
- The orientation of the rock fabric within the slope enhances the potential for slope failure. Shale is an especially soft and weak sedimentary rock. Slope failure is likely to occur along the shale layer which dips parallel to the slope. Even the foliation within the gneissic bedrock parallels the hillslope. Planes of weakness within a rock typically occur parallel to the rock's textural fabric.
- If the slope was undercutting by the road builders, this will also compromise the slope stability.

B. Given that the home is already built on this site, briefly discuss two possible ways of reducing the risk of damage to the house due to slope failure.

This is an inherently unstable slope. A slope ordinance probably should have restricted building on this slope. However, given that the house is already there, what might be done to decrease the risk for slope failure? Reasonable approaches include:

- Don't water the lawn or replace it with a ground cover that requires no watering and sends down deep roots.
- Drain water from the slope above and off the roof away from the gravel fill beneath the house.
- At some expense, rock bolts could be installed to help stabilize rock layers.
- Maintain a good cover of vegetation on the slope above and below the house.
- Put in a retaining wall along the road below the house and be sure that wall does not restrict the drainage of water out of the slope.

Answers to Review Questions

1. A. Site D may have the best view of the shoreline and Site B the best overall view because of it's position on top of the hill, but Site A is on the most stable ground.
2. B. Mass movements occur when the force of gravity exceeds the strength of the slope materials.
3. A. Undercutting by a river or waves will oversteepen a hillslope and enhance the potential for slope failure. Because water can act as a lubricant and also adds weight to the slope materials, draining it will reduce the weight of the slope material and increase friction, thereby reducing the potential for slope failure.
4. C. Talus refers to the blocks of rock that collect at the base of a steep slope or cliff.
5. D. The angle of repose for most loose sands is about 35°. The angle of repose varies significantly with a number of factors, one of which is the size of the particles (refer to Figure Story 12.1).
6. B. Bedding, joint planes, or a foliation fabric all are potential planes of weakness within rock. The orientation of any of these fabrics parallel to the hillslope compromises slope stability (refer to Figure Story 12.1).
7. C. Refer to Figure Story 12.1.
8. A. Solifluction is a result of repeated freezing and thawing.
9. C. Draining the water from the landslide area would help to reduce the weight of the slope materials and reduce the potential lubricating effects of water.
10. D. Bedding planes, joint planes, or textural fabrics, such as foliation, are zones of weakness within rock. If they parallel the hillslope, the potential for slope failure is enhanced.
11. B. Refer to *Unconsolidated Mass Movements* in your textbook.
12. D. The barren slopes left by the wildfires will enhance the potential for all of the hazards listed in answers A–C.

Chapter 13

Answers to Practice Exercises

Exercise 1: Evaluating Rock Materials as Potential Aquifers

Rock material	Porosity (high, medium, low)	Potential as an aquifer (good, moderate, poor)
Loose, well-sorted, coarse sand	High	Good
Silt and clay	*Low*	Poor
Granite and gneiss	Low–Interlocking grains of silicate minerals provide for little pore space.	*Poor*
Highly fractured granite	Medium–Fracturing can significantly increase pore space and improve permeability.	Moderate
Sandstone	*Medium*–The cement that holds the sand grains together reduces pore space. Nevertheless, sandstones are typically good aquifers.	Moderate to good
Shale	Low–Fracturing will increase pore space but permeability may still remain low. Shales are typically aquicludes.	Poor
Highly jointed limestone	Medium–Fracturing and the formation of a cavern system within the limestone can greatly enhance the porosity and permeability of limestones. Caverns serve as an open plumbing system for groundwater.	*Moderate to good*

Exercise 2: Evaluating Groundwater Wells

Well A — potential for:

1. Pollution – High because the water table slopes towards Well A and up slope is the outhouse. The cone of depression produced by pumping Well A enhances the potential for pollution by creating a larger gradient in the water table between Well A and the outhouse.

2. Artesian flow – None: Well A will exhibit no artesian flow because the aquifer is not confined nor sloping.

3. Discharge – High because the aquifer is a porous and permeable sandstone of large volume and Well A is drilled deep into the aquifer.

Well B — potential for:

4. Pollution – Assuming the house does not release pollutants, low because the water table slopes away from Well B and the outhouse is down gradient. The cone of depression around Well B is small and not an issue.

5. Artesian flow – None: Well B will exhibit no artesian flow because the aquifer is not confined nor is it sloping.

6. Discharge – As long as the water table does not lower, discharge from Well B will be potentially high.

7. Long-term supply – Low because Well B is shallow, a lowering of the water table, due to a change in climate or due to pumping from Wells A and B at rates that exceed recharge, could compromise the productivity of Well B.

Well C — potential for:

8. Pollution – Low because there is no source of pollution shown for the confined sandstone aquifer.

9. Artesian flow – High because the aquifer for Well C is a tilted and confined layer of sandstone and the recharge area for the aquifer appears to be higher in elevation than the top of the well.

10. Discharge – High because with high mountains to the east there should be good recharge of this artesian aquifer. The shale layers above and below the sandstone confine the porous and permeable sandstone and, thereby, allow for the development of significant water pressure within the aquifer. Refer to Figure 13.10.

Answers to Review Questions

1. B. Evaporation from the oceans is more than 6 times more than from the land surface. Refer to Figure 13.2.

2. D. Polar ice caps and glaciers contain about 2.97% of all water on Earth. This is the second largest reservoir for water. Refer to Figure 13.1.

3. A. Refer to Figure 13.1.

4. C. Rock particles typically can compact more efficiently as grain size decreases. This reduces porosity and permeability. Refer to Figure 13.7 and Table 13.2.

5. B. Refer to Figure 13.8.

6. C. Permeability is the ability of a solid to allow fluids to pass through it. Generally, permeability increases as porosity increases, but not always. Permeability also depends on the sizes of the pores, how well they are connected, and how tortuous a path the water must travel to pass through the material.

7. A. Stalactite holds "tight" to the ceiling.

> **Memory Tip**
>
> You can easily remember the difference between a stalactite and a stalagmite because stalactites hold "tight" to the ceiling and stalagmites "might" reach the ceiling of a cave.

8. C. Carbonate rocks, such as limestone, are susceptible to dissolution. A wet climate favors cave formation.

9. D. An aquifer is a rock that will yield a good flow of groundwater.

10. C. Refer to the section titled *Water Deep in the Crust* in Chapter 13.

11. D. Shales are typically aquicludes.

12. D. Porosity and permeability both influence how much water an aquifer can produce. Porosity represents the total amount of pore space in which groundwater can be stored. Permeability is linked to porosity and represents the ability of water to flow through the rock. Some shales and lava flows have moderate to high porosity but very low permeability because the pore spaces are not interconnected. Some rocks, such as granite, are moderately permeable because of extensive fracturing. However, their porosity may be low because the fracturing is very tight.

13. B. Loose sand will typically exhibit high porosity and permeability because the round sand grains can not pack together very efficiently (refer to Figure 13.7).

14. A. Refer to Figure 13.11.

15. B. Rivers are fed by springs from a shallow groundwater table that intersects the stream channel. Refer to Figure 13.11.

16. A. Refer to Figure 13.10.

17. D. Well D. Refer to Figure 13.10.

18. B. Sand and gravel will typically exhibit a high porosity and permeability.

19. D. Sandstones are typically good aquifers. Because water flows down hill, it will tend to collect and recharge aquifers located beneath topographically low spots, such as valleys.

20. B. Well B. The cone of depression around well B has lowered the water table enough that effluent leached from the outhouse will flow towards well B, even though the topography slopes toward the river. Refer to Figures 13.12 and 13.20.

21. B. Well B will be the most productive water well because it taps into a larger sandstone aquifer. Well A is not likely to be productive because it bottoms out in a shale which typically has very low permeability. Well C will yield water because it bottoms out in the sandstone, above a shale unit that creates a perched water table. The volume of water from well C will be limited.

22. C. Refer to Figure 13.14.

23. A. The spring is least likely to be contaminated by the outhouse because it is produced from a perched water table well above and independent of the aquifer that is affected by the pollutants from the outhouse.

24. C. h/l for A = 0.04, B = 0.03, C = 0.05, and D = 0.006.

Chapter 14

Answers to Practice Exercises

Exercise 1: Stream Velocity

Variable affecting stream velocity	Relationship of variable to stream velocity	Analogy
Gradient–the slope of the stream channel	As gradient decreases, stream velocity decreases. Velocity is proportional to the gradient. Note: In headwaters of streams (in the mountains where gradient is highest) other factors (decreased discharge and increased channel roughness) will counter the effect of the high gradient.	Do you tend to walk faster down a steeper slope or a more gradual slope?—faster down a steeper slope.
Discharge–the amount of water in the stream channel	As discharge increases, stream velocity increases. Velocity is proportional to the discharge. Note: Surprisingly large objects can move down a river during a flood.	Will you move into a new house slower or faster if you have more people helping you?—faster.
Sediment load	As the availability of sediment increases, the stream velocity will decrease. Note: Various factors, such as the bedrock, a landslide, erosion of soil from a burned area, and construction, can influence the availability of sediment load.	Typically, will you travel faster or slower if you are carrying a heavier load?—slower.
Channel characteristics:		
• Channel roughness	*As channel roughness increases, velocity will decrease.*	*Cross-country hiking without a trail tends to slow one down.*
• Channel shape	*The stream has more contact with the channel surface if the channel shape is very wide or very narrow. More contact with the channel will increase drag and decrease velocity.*	*When you have more contact with the ground surface (for instance, when you crawl on your knees), you move slower than when you have less contact with the ground surface (for instance, when you are walking upright on your feet).*

Exercise 2: Relationship between Stream Flow and Groundwater

Why do streams in desert regions typically flow intermittently and streams in more temperate regions, such as New England, flow year-round?

Effluent streams are fed by springs from a water table that intersects the stream channel. The water table is well below the channel of an influent stream which flows in response to rainfall but will quickly dry up as runoff from a storm decreases. Infiltration of stream flow into the channel bottom sediments may help to recharge the groundwater table beneath the channel.

Exercise 3: How Do Rivers Cut through Mountain Ranges?

The Kali Gandaki River cuts one of the deepest gorges on Earth right through the Himalayan Mountains. Briefly describe two ways that a river can cut through a mountain range. Refer to Figures 14.21 and 14.22.

Two ways a river can cut through a topography obstruction, such as a mountain range, are:

A. **Antecedence** where the river existed before the present topography was created, and it maintained its original course despite changes in the underlying rocks and topography and;

B. **Superposition** where the river was established at a higher level, on a uniform surface before eroding down and superimposing itself on a buried geologic structure, such as an anticline.

Answers to Review Questions

1. B. Gravity is the force that drives water down hill.
2. B. Discharge is the volume (typically cubic feet or meters) of water flowing past a point on the stream channel for a given interval of time (typically seconds).
3. A. Velocity is directly proportional to gradient. An increase in slope will increase velocity if all other factors do not counter the change in slope.
4. B. Stream velocity is the most important variable determining the behavior of a stream.
5. A. The change in elevation was 200 feet over the 200 miles canoed. 200 feet/200 miles = 1 foot per mile.
6. B. Refer to Figure 14.20c.
7. A. Stream competency is a measure of the size of particles a stream can transport. Stream capacity is a measure of the amount of sediment load a stream can transport.
8. B. Coarser particles, such as pebbles, will settle out before sand and silt.
9. B. The outer bank around a meander is much more likely to be eroded because stream velocity is fastest around the outside of the meandering channel where water depth is greater. On the inside of a meander water depth is lower, velocity slows, and the stream is more likely to deposit sediment and form a point bar. Refer to Figure Story 14.9.
10. C. An increase in rainfall, in stream gradient, or a lowering in base level will increase the stream's velocity and give the stream renewed ability to downcut. Typically, the stream will entrench its channel along its preexisting meandering course.
11. C. Refer to Figure 14.15.
12. C. Regular addition of sediment load is likely to cause the stream velocity in the disturbed stretch to decrease due to the increased availability of load and a reduced gradient. A reduction in velocity may induce deposition.
13. D. The net effect of straightening out the river channel by cutting off a meander bend is to increase the gradient of the stream channel in the cutoff section. An increase in gradient will result in an increase in velocity and increased potential for channel erosion.
14. D. A 50-year flood event has a 2% chance of occurring any year over a 50 year period.
15. D. This may surprise you but stream velocity typically increases downstream. The reduction of stream gradient is countered by the increase in discharge from tributary channels. The lazy old Mississippi River runs faster past Vicksburg, Mississippi, than up at St. Louis, Missouri.
16. C. A lowering of base level increases the stream gradient giving the stream renewed ability to erode and transport rock material. Erosion will begin in the vicinity of the change.
17. B. Refer to Figure 14.16 and associated text. Answers A and C are incorrectly stated because velocity would decrease with a decrease in gradient and an increase in discharge would result in erosion not deposition.
18. D. Choices A–C are all possible results of regional uplift and would raise the headwaters of the drainage system, which would increase stream gradient and enhance the potential for stream erosion. See Figures 14.8 and 14.17.
19. A. House A is located on the outside of a meander bend which is most susceptible to bank erosion. Refer to Figure Story 14.9.

Chapter 15

Answers to Practice Exercises

Exercise 1: Sand Dune Types

Dune type	Characteristics	Sand supply	Wind direction/strength
Barchan	Crescent-shaped with arms pointing downwind and slip face is concave curve advancing downwind	Limited	Unidirectional/strong
Transverse	Long, wavy ridges oriented at right angles to the prevailing wind	Abundant	Unidirectional/strong
Blowout	Crescent-shaped with arms pointing upwind (into the wind) and slip face forms convex curve advancing downwind	*Limited to moderate*	*Unidirectional/gusty*
Linear See Figure 15.8	Long, straight ridges more or less parallel to general direction of the wind	Moderate	Variably direction/moderate to strong

Answers to Review Questions

1. D. Refer to Figure 15.1 and the section in the text titled *Wind Belts*.
2. A. Rising air is characteristic of equatorial latitudes where the sun's radiation is more concentrated. As the air rises, it cools and releases abundant rainfall typical of tropical regions. The other three answers all contribute to desert conditions.
3. D. Refer to Figure Story 15.11.
4. D. Refer to Figure Story 15.11 and the section in the text titled *How Sand Dunes Form and Move*.
5. C. Refer to Figure 15.18. Because it is so flat and smooth, a playa (dried lake bed) in the Mojave Desert east of Los Angeles is a regular landing site for the Space Shuttle.
6. A. Loess is fine wind-blown dust.
7. B. Refer to Figure 15.7.
8. B. Silt and clay are carried in suspension. Sand is typically transported by saltation, sliding, and rolling.
9. B. Refer to Figure 15.1 and the section titled *Wind Belts* in the text.
10. D. Direction D. The arms of the barchan point downwind. Because the wind is blowing from south to north, the arms are pointing north and the town is in the opposite direction (D).
11. B (South). The arms of blowout dunes point into the wind (upwind). Because the wind blows inland from the coast, the arms of the dune point to the beach.
12. D. All of the above answers are appropriate.
13. Pedocal, calcium carbonate. Refer to Figure 7.16.
14. Desert varnish. Refer to Figure 15.16.

Chapter 16

Answers to Practice Exercises

Exercise 1: The Glacial Sculptured Landscape

The section in the text titled *Glacial Landscapes* and Figures 16.19, 16.20, and 16.21 will be very helpful for completing the brief descriptions of glacial features. A list of possible glacial features that might be found and interpreted by a seasonal ranger include:

Glacial features formed by the erosive power of glacial ice

- striations—scratches and grooves—carved in bedrock over which the glacier flowed.
- cirque
- U-shaped valley
- hanging valley
- fjord
- arête

Glacial features formed by deposition of rock material by glacial ice

- glacial moraines—the different types are described in Table 16.1.
- drumlins
- esker
- kame
- glacial erratic
- kettle
- varves

All of the above glacial features are described in your textbook and are appropriate answers to Exercise 1.

Exercise 2: Your Personal Budget as a Metaphor for a Glacial Budget

See the section of your text titled *Glacial Budgets: Accumulation Minus Ablation*.

A personal checkbook is a good metaphor for the dynamic balance between accumulation and ablation on a glacier. For example, if you deposit money into your checking account faster than the rate of withdrawal, the cash balance grows. In a similar fashion, as snow accumulates faster than the rate of ablation (loss), the glacier expands and advances. If you withdraw money from your checking faster than you deposit it, you account shrinks. Likewise for a glacier, if the rate ablation exceeds the rate of accumulation the glacier will shrink and retreat up slope. It might disappear all together if ablation exceeds accumulation for an extended period.

Exercise 3: Glacial Advances and Retreats

Rarely does a glacier actually remain stationary. Driven by the force of gravity, glacial ice and the rock material that it carries are moving down hill. The words advancing, retreated, and halted are used to describe the movement, or location, of the toe or terminus of the glacier and do not actually refer to the movement of glacial ice within the glacier. The terminus of the glacier will remain stationary (halted), retreat up the valley, or advance down the valley depending on the glacial budget. Refer to Figure 16.10. For example, if the snow accumulating in the upper reaches of the glacier equals the loss (ablation) of glacial ice from the lower and warmer reaches of the glacier, the size of the glacier will remain constant and the glacial terminus will remain stationary. Nevertheless, the glacial ice is still flowing down slope with rock material and may pile up a sizeable end moraine. Refer to Table 16.1.

Answers to Review Questions

1. C. Glacial ice best fits the general definition of a metamorphic rock. Refer to *Ice as a Rock* in your textbook.
2. D. Gravity is the force that pulls glacial ice down hill. In response to gravity glacial ice moves by plastic flow and basal slip.
3. B. Refer to Figure 16.8.
4. A. Refer to Figure 16.10.
5. D. Moraines are deposits of till. Refer to the section titled *Glacial Sedimentation and Sedimentary Landforms* and Table 16.1 in your text.
6. D. Drumlins are streamlined hills of till and bedrock. Because drumlins parallel the direction of ice movement, they can be used to reconstruct the direction of movement for the ice sheet.
7. D. Arête.
8. A. Water is tied up as glacial ice on land. Therefore, sea level drops.
9. C. Refer to Earth Issue Box 16.2.
10. D. Refer to Figure 16.27.
11. C. Northern Europe received up to 40% of its total annual heat from the Gulf Stream. Refer to Figure 16.28.
12. B. Refer to 16.27.
13. B. Kames are small hills of sand and gravel dumped near or at the edge of the ice. Refer to Figure 16.21.
14. A. Because glacial ice is capable of carrying rock particles of a great variety of sizes and physical weathering dominates in glacial environments, it is very unlikely to get a glacial till consisting of pure quartz sand.
15. A.
16. A. Refer to Figure 16.10.
17. C. Refer to Figure 16.24.
18. D. All of the factors listed play a role in climate change and need to be considered as components of a model that attempt to describe the cause of the ice ages.
19. D. The mass of the floating ice is equal to the mass of the water the iceberg displaces. When the ice melts, it simply replaces the water it displaces and, therefore, there is no change in sea level.

CHAPTER 17

Answers to Practice Exercises

Exercise 1: Profile from the Atlantic Shoreline to the Ocean Floor

A. Fill in the blanks to correctly label the following profile. **Hint:** Refer to Figure 17.8a.

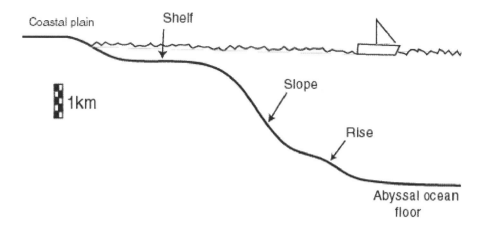

B. Does the profile above illustrate an active or passive continental margin? Explain. The profile characterizes a passive continental margin. Refer to Figures 17.8 and 17.9 and the associated text.

Exercise 2: Passive vs. Active Continental Margins

A passive margin is a continental borderland far from an active plate boundary. In contrast, active margins are associated with subduction zones and transform faults. The volcanic activity and frequent earthquakes give these narrow and tectonically deformed continental margins their name.

Continental shelves are broad and relatively flat at passive continental margins and are narrow and uneven at active margins.

A. passive

B. passive

C. active (subduction/convergent plate boundary)

D. passive

E. active

F. passive

Answers to Review Questions

1. C. The wind generates most waves in the oceans.
2. C. The edge of the continent is rarely if ever the shoreline. It is the continental slope and rise, which may be hundreds of kilometers from the shoreline.
3. A. Mafic volcanism at the oceanic spreading centers generates new ocean crust.
4. D. Wave refraction is the tendency of a wave to bend into the shoreline that it approaches. Waves bend as the wave bottom "drags" on the shallowing bottom before the shoreline, slowing that portion of the wave down. See Figure Story 17.13.
5. A. Refer to Figure 17.12d and pages 398–399.
6. A. Mafic volcanism at divergent plate boundaries generates the basaltic seafloor crust, which is over time buried in pelagic sediments.
7. A. Seamounts are extinct submarine volcanoes made mostly of basalt.
8. B. Refer to Figure 17.12d.

9. D. 4000 to 6000 meters is the depth of the abyssal ocean floor.

10. A. Sea level appears to be rising about 4 millimeters per year.

11. A. Turbidity currents plunge down the continental slopes often triggered by earthquakes. They are flowing masses of turbid muds suspended in water, denser than the clear water above it. See Figures 17.9.

12. A. Refer to Figure 17.14.

13. C. The groin will block the drift of sand past your beach in the surf zone. Robbed of a continuous supply of sand, longshore currents will carry sand from your property, which will not be replenished by the sand supply up current. Your sandy beach is likely to disappear in a few years.

14. B. Wave energy is focused on points and headlands by wave refraction.

15. C. Refer to Figure 17.18 illustrating the sand budget—the dynamic balance between inputs and outputs of sand along the shoreline.

16. A. Rip currents are generated by rapidly moving backflows and run perpendicular to the beach. Therefore, the best way to escape from the grip of a rip tide is to swim parallel to the beach, unless you are up for some long distance swimming.

17. C. Reef forming corals require sunlight to grow and do not thrive in water much deeper than 20 meters. Therefore, atolls begin as fringing reefs around a volcanic island. Over time as the volcanic island slowly subsides, the coral reef grows upward, maintaining a shallow marine environment. Refer to Figure Story 8.16 and Earth Policy Box 8.1.

18. C. Refer to Figure Story 17.13.

19. A. Refer to Figure Story 17.13.

20. C. Turbidity currents (flows) are both agents of erosion and deposition of sediment along the continental slope and adjacent ocean floor. Refer to Figure 17.9.

21. The longshore current will flow parallel to the beach in the surf zone and to the left, since the waves are coming in at an angle from the lower right. The best answer has the arrow close to the beachfront in the surface zone. See Figure Story 17.13.

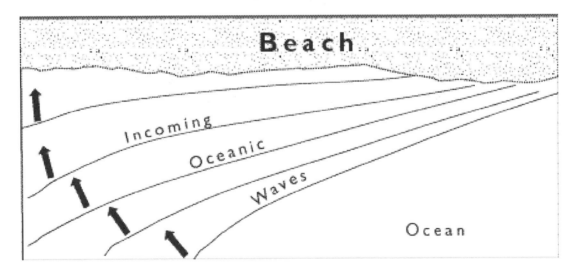

22. A. Refer

Chapter 18

Answers to Practice Exercises

Exercise 1: Landscapes: Tectonic and Climate Interaction Flowchart

To review the basic relationships between landscape relief, tectonic activity (uplift), and erosion, fill in the flowchart below. Place the following words in the correct positions on the flowchart:

- tectonic activity
- high relief
- low relief
- erosion
- physical weathering
- chemical weathering

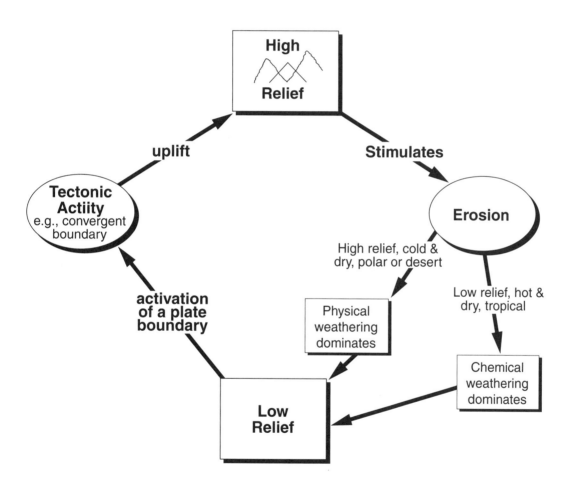

Exercise 2: Comparison of Some of the Landforms

Landform	Important feature(s)	Sketch (Hint: Keep it very simple)
Mesa See Figure 18.8.	A small plateau with steep slopes on all sides. Held up by underformed Sedimentary layers or lava flows.	*sketch of Mesa*
Cuesta See Figure 18.14.	A structurally controlled cliff. Some what tilted beds with alternating weak and strong resistance to erosion. Typically undercut and asymmetrical.	*sketch showing S.S. layers*
Hogback See Figure 18.15.	A structurally controlled cliff with beds that are steeply dipping to vertical. Ridge is more or less symmetrical.	*sketches: Side view and end view*
Valley ridge topography See Figures 18.10 and 18.11.	In young mountains, upfolds (anticlines) form ridges and downfolds (synclines) form valleys. As tectonic activity moderates and erosion digs deeper into the structures, the anticlines may form valleys and syncline ridges.	*sketch of Ridge-Valley-Ridge with anticline and syncline*

Answers to Review Questions

1. A. River valleys begin with a V-shaped profile. As the sides of the steep valley retreat and the floor of the valley widens, a floodplain and relatively flat-floored valley will evolve. The gradient in a youthful river system is steep but decreases though time unless other geologic events are superimposed on the history of the drainage. Glaciers carve U-shaped valleys. Refer to Figure Story 18.12.

2. B. Elevation is the result of the balance between tectonic activity (uplift and subsidence) and erosion.

3. A. Refer to Figure 18.3.

4. D. Tectonic activity, erosion, climate, and type of bedrock are important controls on landscape evolution.

5. D. Earth's surface has two fundamental levels, the continents, which on average are about a half a mile above sea level, and the ocean floor, which on average is about 2.5 miles below sea level.

6. B. Although mountains form at all four of these tectonic settings, the longest and highest mountain ranges form at convergent plate boundaries.

7. A. Refer to Figure 18.17 and pages 417–419.

8. D. Refer to Figures 18.19 and 18.11.

9. A. Refer to Figure 18.12.

10. B. Refer to Figure 18.15.

11. A. Refer to *Earth Issues* box 18.1.

12. D. See page 423.

CHAPTER 19

Answers to Practice Exercises

Exercise 1: Earthquake Focus vs. Epicenter

What type of fault is this? normal fault

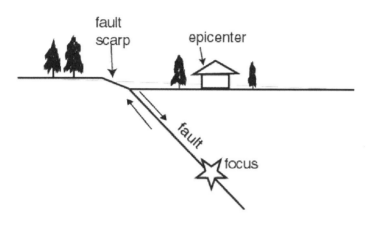

Exercise 2: Characteristics of Seismic Waves

Figure Story 19.5 and the section of the textbook on *Seismic Waves* will be helpful.

Seismic waves	P (primary) waves	S (secondary) waves	Surface waves
Relative speed	Fastest	*Second fastest*	Slowest
Motion of material through which wave propagates	Compressional—push/pull	Shearing	*Rolling/elliptical and sideways motions*
Medium through which wave will propagate	Solids Liquid Gas	Solids only	*Confined to the Earth's surface*
Analogy with common wave forms	Sound waves	*S waves propagation is difficult to visualize. It is somewhat analogous to the way cards in a deck of playing cards slide over each other as you shuffle the deck.*	Ocean waves

Exercise 3: Factors that Amplify the Damage Caused by an Earthquake

Factors that influence the damage caused by an earthquake include:

- The magnitude of the earthquake. Obviously, a larger magnitude earthquake tends to be more destructive but many other factors beside magnitude can contribute to the potential for destruction.

- The duration of the earthquake. Earthquakes that last longer usually cause more damage. Consider this analogy. The longer you push someone on a swing, the greater the swing back and forward. Likewise, if structures experience a longer duration of pushes and pulls by seismic waves, the more they will sway and the greater the potential for damage.
- The ground acceleration caused by the earthquake. Faster ground acceleration is likely to cause more damage. Rapid ground acceleration may have contributed to the destructiveness of the Northridge, California, earthquake in 1994.
- The depth to the focus of the earthquake. The energy of seismic waves dissipates with distance from the focus.
- Proximity to a coastline where a tsunami may hit. Refer to *Earth Issue* box 19.1.
- Fires ignited by ruptured gas and power lines. Fire started during the 1906 San Francisco earthquake caused by far the most damage to the city.
- Liquefaction of water-rich soils and sediments that are typically associated with coastal regions and developments. Liquefaction was a big factor in the damage caused to Anchorage during the 1964 earthquake.
- Preexisting building designs and codes for the community.
- Landslides and unstable slopes.

You may come up with other reasonable factors that are not listed above.

Answers to Review Questions

1. B. Refer to Figure Story 19.1.
2. B. An earthquake propagates from the focus of the earthquake. The epicenter is the location on the surface directly above the focus.
3. A. P waves travel fastest and surface waves travel with the slowest velocities.
4. C. Three. Refer to Figure Story 19.6.
5. C. Richter Scale measures the amount of ground motion. Refer to Figure 19.7.
6. D. The Richter Scale for earthquake magnitude is exponential. With every unit of increase in magnitude, like from 4 magnitude to 5 magnitude, the ground motion increases by a factor of 10.
7. C. P waves, like sound waves, travel through solids, liquids, and gases.
8. B. S waves travel only though solids.
9. B. Earthquakes are strongly associated with all active plate tectonic boundaries.
10. B. Producing a seismic risk map would be a first step in earthquake risk assessment.
11. D. Water-saturated stream delta sediments exhibit liquefaction in response to the shaking from an earthquake.
12. C. 3000 kilometers based on the seismic travel-time curves in Figure 19.6.
13. C. Deep focus earthquakes do not occur in association with divergent boundaries. How do you explain this observation?
14. C. Texas has the lowest seismic risk compared to the other regions listed. See Figure 19.19.
15. B. Refer to illustration to the right. Notice that the arrows are pointing west and east to reflect the pushing and pulling of the land.

16. This would be a left-lateral strike-slip fault. Refer to illustration to the right.

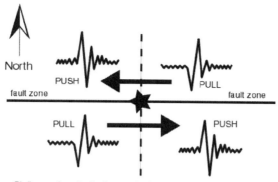

CHAPTER 20

Answers to Practice Exercises

Exercise 1: Evolution of the Continents

Complete the sentences in the flowchart by filling in the blanks with words from the list. Refer to Figures 20.12, 20.15, and 20.18.

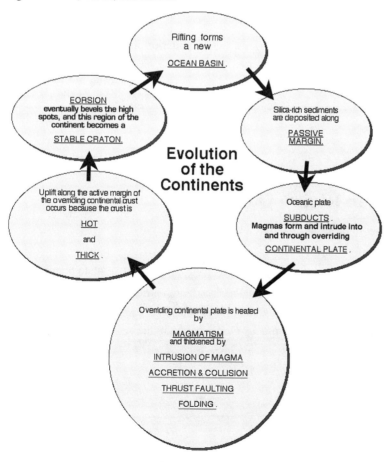

Exercise 2: Ocean Crust vs. Continental Crust

Using information in Chapters 2, 5, 17, 18, and 20, complete the following table by filling in the blank boxes.

Characteristics	Ocean crust	Continental crust
Composition	Mafic	Felsic to intermediate
Rock type(s)	Homogeneous basalt, gabbro, and pelagic sediments	Very heterogeneous—can contain any rock but dominantly granitic and gneissic with a cover of sediments
Density	3.0 g/cm^3	2.7 g/cm^3
Thickness	10 km	15–65 km
Age	175 million years or younger with older fragments caught up within continents	The ages of continental crust spans 4 billion years.
Topographic features	Abyssal floor Ridge with axial rift Trenches Seamounts Hot spot island chains Plateaus	Craton or shield Stable platform Continental margin Coastal plain Continental shelf and slope Mountain belts Sedimentary basins
Structure/ Architecture	A model for the structure of the ocean crust is the ophiolite suite: deep-sea sediments, basaltic pillow lavas and dikes, and gabbro. (Note: peridotites are part of the mantle lithosphere, not the ocean crust.)	The architecture of the continents is complex. It consists of preexisting cratons, accreted microplates, island arcs, volcanic arcs, suture zones, ophiolite suites, and belts representing ancient orogenic zones. Sediments cover basement rock in the interior platform of the continent.
Origin	Mafic magmatism and volcanism at the ocean ridge system	Orogenic processes and accretion of preexisting crustal blocks along convergent plate boundaries

Answers to Review Questions

1. C. Refer to pages 476–480 in the text.
2. A. Refer to Figure 20.12.
3. C. Passive continental margins are located far from active plate boundaries.
4. C. The oldest known continental rock is radiometrically dated at 4 billion years old. It is the Acasta gneiss from the Slave Province in Canada. Refer to Figure 20.22.
5. D. Refer to the *How Continents Grow* section in the text.
6. C. Refer to Figure 20.18.
7. A. Refer to Figure 20.11.

8. B. Refer to Figures 20.1 and 20.8
9. C. Young orogenic belts are characterized by hotter and thicker crust.
10. A. Refer to Figure 20.11 and *The North American Cordilleran* section in your textbook.
11. B. The Andes Mountains along the west coast of South America are an active orogenic system associated with the subduction of ocean crust beneath the South American continent.
12. A. Orogeny refers to the collection of processes, typically active at convergent plate boundaries, which form mountains. Rifting leads to the formation of a new ocean basin.
13. C. Refer to *The Appalachian Fold Belt* section in your textbook.
14. B. See *The North American Cordillera* section in your textbook.
15. A. The Cordillera is a younger mountain belt compared to the Appalachians. In fact, portions of the Cordillera are still involved in orogenic processes. Western North America crust sits higher than the Appalachians because the Cordillera crust is thicker and hotter, and tectonic processes remain very active.
16. D. Widespread and relatively thick accumulations of coral-rich limestones, sandstones, and shale is characteristic of a passive continental margin or a sedimentary basin within a continental platform.
17. A. Thicker continental crust typically stands higher. Hotter crust will also sit higher because heat lowers the density of the rock material.
18. B. In a sense, mountains are made from mountains. Most of the sediments eroded from previous orogenic systems ultimately end up along a continental margin where they eventually will be deformed and uplifted by the evolution of an active convergent boundary. Some sediments are probably subducted, but most are deformed and entangled in thrusting and metamorphism during orogeny. Some sediments may be carried to depths where they melt. Melts derived from sediments are typically felsic, not mafic, and crystallize within the crust to form granitic rocks. Except for regions adjacent to the continental margin, most ocean floor is surprisingly lacking in sediment. This is partly because the ocean floor is young and continually being recycled by plate tectonics and because sedimentation rates in the deep ocean are very slow.
19. B. Rifting breaks up of a continent and forms a new passive margin on which sediments accumulate. Eventually with the development of subduction adjacent to the passive margin, orogeny is initiated. Orogenic processes thicken the overriding crust, which then begins to rise due to a buoyant balance between the mountain root and the denser mantle in which it sits. Can you think of a good reason why subduction is likely to eventually occur adjacent to an old passive continental margin?
20. C. Refer to Figure 20.8.
21. B. Refer to the *How Continents Grow* section.
22. A. The Andes Mountains along the west coast of South America are in a more youthful stage of orogeny than the North America Cordillera and exhibits a great number of similarities to past geologic circumstances in western North America. Today geologists actively study the Andes as a way of time traveling back into the earlier Cenozoic history of the North American Cordillera.
23. B. Refer to Figure 20.20.
24. A. Refer to Figure 20.24.
25. D. Refer to Figure 20.6.

Chapter 21

Answers to Practice Exercises

Exercise 1: The Earth's Interior Layers

Fill in the blanks. Use Figures 21.5, 21.6, and 21.7 as references, in addition to the section The Layering and Composition of the Interior in this chapter.

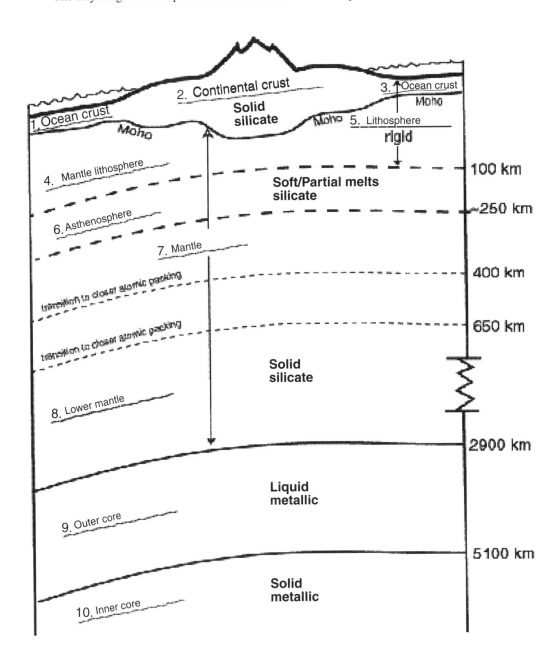

Exercise 2: The Characteristics of Earth's Internal Layers

Complete the following table by filling in the blanks and completing the sentences. Shaded boxes remain blank. Refer to the text and figures in Chapters 1, 2, 5, 20, and 21.

Answers to Practice Exercises and Review Questions 289

Layers	Volume % of total	Mass % of total	Density g/cm³	Physical state	Composition	Observations and evidence that support the characteristics of the layer
Crust	**0.60**	**0.42**				
Continental crust	0.44	0.25	≈ 2.7	Solid	Felsic and intermediate	Very heterogeneous. 40–65 km thick. Formed at convergent boundaries by orogenic processes.
Ocean crust	0.16	0.17	≈ 3.0	Solid	Mafic silicates	Very homogeneous. About 10 km thick. Formed at divergent plate boundaries from partial melting of ultramafic mantle rocks.
Mantle	**83.02**	**67.77**	3.3–5.7		Ultramafic, (peridotite)	
Mantle lithosphere				Rigid, solid	Ultramafic silicates (mostly peridotite)	Crust and mantle lithosphere makes up Earth's tectonic plates. Thickness ranges from 0 km at spreading centers to 200 km beneath continents. S-wave velocities speed up through it.
Asthenosphere				Weak, plastic, solid	Ultramafic silicates (mostly peridotite)	Weak zone. A few percent melting. Reaches close to the surface at spreading centers and deepens under older seafloor. S-waves slow down and are partially absorbed.
Lower mantle				Solid	Ultramafic	Abrupt increase in S-wave velocities at 400 and 650 km mark changes in mantle structure—collapse of the atomic structure of minerals.
Core		**31.79**				
Outer core	15.68		9.9–12.2	Liquid	Mostly iron and nickel	P-waves slow down; S-waves are completely blocked. Iron-nickel composition is consistent with bulk density of the Earth.
Inner core	0.70		12.6–13.0	Solid	Mostly iron and nickel	P waves suddenly speed up at 5100 km. S-waves are transmitted through inner core. Composition also consistent with the natural abundance of iron and meteorites.

Answers to Review Questions

1. C. Earth's core is about one-half the radius of planet Earth. See Figure 21.5: in the illustration, Earth's center is at 6000 km depth, and the outer core is at 3000 km depth.
2. B. Refer to Figure 21.6. The ocean basins exist on Earth because the ocean crust is thinner and more dense than the continental crust. Therefore, the ocean crust sits lower.
3. C. Ultramafic or peridotite.
4. B. Earth's tectonics plates are large fragments of the lithosphere, a rigid layer consisting of the crust and uppermost mantle.
5. C. Continental crust is very heterogeneous in composition. Nevertheless, its average composition is felsic to intermediate.
6. D. The lithosphere is thought to be a rigid layer that rests on a plastic, weak asthenosphere.
7. B. S-waves do not pass through the outer core.
8. C. The metallic core has the highest density.
9. C. Earth's north magnetic pole is located between Greenland and Baffin Island, not at the Earth's pole of rotation—the North Pole.
10. C. The electrodynamo theory explains the Earth's magnetic field as a consequence of electrical currents moving in the fluid, metallic outer core. The silicate minerals in the asthenosphere are not nearly as good conductors of electricity. Therefore, flow of electrical currents and the associated magnetic field are much weaker in the asthenosphere.
11. B. The asthenosphere is thought to be a weak, plastic, zone within the upper mantle that may be partially melted.
12. B. Refer to Figure 21.14
13. A. Refer to Figure 21.9.
14. C. The mantle–core boundary is marked by the most significant change in composition.
15. A. The thickest continental crust on Earth lies beneath the Tibetan Plateau, the Himalaya Mountains, and portions of the Andes Mountains. Refer to Figure 21.6.
16. B. Refer to Figure 21.7.
17. D. Answers A, B, and C are all true.
18. D. If a significant portion of the lower mantle were molten, we should see lots more volcanic activity on the Earth's surface in addition to tectonic plate boundary volcanism.
19. A. Seismic wave velocities depend on the elasticity, rigidity, and density of materials. These properties are dependent on the composition, physical state, and compactness of the atomic structure of the material.
20. C. Refer to Figures 21.11 and 21.13.
21. B. The Moho was first detected by an abrupt increase in seismic wave velocities due to the change from lower density crustal rocks to the more dense ultramafic rocks within the mantle.
22. A. Refer to Figure 21.10 and Earth Issues 21.2.
23. D. Refer to Figure 21.15.
24. A. Refer to Figures 21.1 and 21.2.
25. D. Refer to Figure 21.8.

CHAPTER 22

Answers to Review Questions

1. B. Refer to Figure 22.8.
2. C. Uranium is not a fossil fuel. Refer to Figure 22.4 for how energy is stored in fossil fuels.
3. B. Ozone depletion in our atmosphere is largely influenced by the release of chlorofluorohydrocarbons (CFCs), synthetic chorine, and fluorine compounds used widely in aerosols and refrigerants through much of the last century.
4. B. Oil and gas form from organic matter derived from marine organisms that thrive in shallow coastal waters.
5. D. Refer to Figure 22.5.
6. A. In 1859 the first oil well in America was drilled in northwestern Pennsylvania. The oil from this shallow well was sold for dubious medicinal purposes.
7. C. Refer to Figure 22.3.
8. C. There are 42 gallons to a barrel of oil, of which about half is refined into gasoline for automobiles. From the remaining oil, jet fuel, diesel and other fuels, solvents, lubricants, greases, and asphalt are produced.
9. C. The United States ranks eighth in oil reserves. See the section *The World Distribution of Oil and Natural Gas*.
10. D. The former Soviet Union has the greatest coal reserves. Other leading producers are the United States and China.
11. B. Refer to Figure 22.8.
12. B. Refer to Figure 22.8.
13. B. Sulfur dioxide combines with rainwater to form sulfuric acid, a major component of acid rain. For this reason, low-sulfur coals are environmentally more favorable to burn.
14. A. Refer to the *Sedimentary Mineral Deposits* section in your textbook.
15. C. Metallic ore deposits have a strong association with divergent and convergent plate boundaries. Refer to Figure 22.23.
16. B. The formation of copper ore deposits is strongly linked to hydrothermal and magmatic processes. Refer to the *Disseminated Deposits* section in your textbook.
17. D. Kimberlites are the host rock for diamonds. Because of their resistance to weathering, diamonds also occur as placer deposits. Diamonds are formed naturally by the extreme pressures of the upper mantle; thus their association with kimberlite, a type of peridotite.
18. B. Refer to the *Nonmetallic Sedimentary Deposits* section in your textbook.
19. D. Metamorphism associated with magma intrusion would probably completely decompose fluid hydrocarbons. Oil and gas would not be associated with volcanic rocks and gneisses. Only very unusual circumstances might result in the fractured gneiss being a reservoir rock for oil and gas. Mesozoic sand dune deposits might make a good reservoir rock but are definitely not source rocks for the hydrocarbons. Oil and gas are derived from the transformation of abundant organic matter deposited in shallow marine sediments. Low-angle thrust faults do produce geologic traps for oil and gas in the western United States.

20. A. Refer to the *Hydrothermal Deposits* section in your textbook.

21. B. Well B is most likely to produce oil because it is drilled into a tilted sand layer, which is confined by the salt dome. Hydrocarbons are likely to accumulate in sand because of its high porosity and permeability. Because salt deposits are essentially nonporous and impermeable, the dome would seal off the sand layer and prevent the fluid hydrocarbons from escaping to the surface. Well A is drilled in shale, which can be a good source rock for hydrocarbons, but its low permeability makes it a very poor reservoir rock. Well D is drilled into a sand layer that is not sealed off by the salt dome. Therefore, hydrocarbons would migrate along the Well D sand layer to the surface and be lost.

22. A. Refer to Figure 22.4.

23. B. Refer to *Earth Policy* box 22.1.

24. B. Refer to Table 22.1.

CHAPTER 23

Answers to Practice Exercises

Exercise 1: Conceptual Map/Flowchart of a Climate Factor

Climate is a complex system with many factors potentially influencing it. Potential focuses for your conceptual flowchart include

High mountains → rain shadow effect

El Niño → wet weather in the eastern equatorial Pacific

Gulf Stream → transfers tremendous amount of heat to northern Europe

Earth's orbital (Milankovitch) characteristics → cyclical changes in the amount of energy reaching Earth.

Volcanic eruption injects dust and aerosols into atmosphere → increase in albedo → Earth cools

Volcanic eruption releases tremendous amounts of carbon dioxide gas → enhanced greenhouse effect → Earth's surface heats up.

Many other factors could be used as answers for this exercise.

Exercise 2: Release of Carbon Dioxide from Burning of Fossil Fuels

A. Explain why the release of carbon dioxide from the burning of fossil fuels might lead to an increase in cloud cover.

An increase in CO_2 enhances the greenhouse effect and results in warmer surface conditions. Warmer surface conditions increase evaporation rates and more water vapor goes into the atmosphere. More water vapor in the atmosphere may result in more cloud cover.

B. Would an increase in cloud cover represent positive or negative feedback? Explain. For example, would an increase in cloud cover enhance or reduce the effect of increasing carbon dioxide in the atmosphere? See pages 547–549 in the textbook for examples of feedback mechanisms that can help to balance or stabilize the climate system.

An increase in cloud cover may produce negative feedback because cloud cover typically has a cooling effect on surface conditions. So the impact of increased cloud cover is opposite to that of increased CO_2 in the atmosphere.

How hard to realize that every camp of men or beast has this glorious starry firmament for a roof! In such places standing alone on the mountain top it is easy to realize that . . . we all dwell in a house of one room—the world with the firmament for its roof—and are sailing the celestial spaces without leaving any track.

—JOHN MUIR, 1890

Exercise 3: Flow of Carbon Through Earth's Systems and Reservoirs

Carbon dioxide is an important greenhouse gas that can greatly impact Earth's surface temperatures. Because the Earth is a closed system, the carbon cycle and distribution of carbon in the various reservoirs is a very important component in any model for how climate changes. **Fill out the following table to summarize the flow of carbon through Earth's systems and reservoirs.** Flux is the amount of energy or matter flowing through a given area or reservoir in a given time. Refer to Figures 23.6, 23.11, and 23.12 and the accompanying text in Chapter 23.

Carbon fluxes	Brief description of flux	Direction of flux	Climatic impact/ implications
Life processes	Carbon is fixed in living organisms, which ultimately contribute to organic matter in sediments, coal, and oil.	Carbon flows from the atmosphere and oceans into rock—the lithosphere.	*Less greenhouse gas in atmosphere results in a cooling.*
Sedimentation	Calcium and carbonate ions combine to produce calcium carbonate, which can precipitate and collect to form limestone or help cement other rock particles.	*Carbon is drawn out to the atmosphere and ocean and precipitated as sediment.*	**Climate cools.** Carbon dioxide is drawn out of the oceans and atmosphere. The loss of CO_2 from the oceans will result in a reduction of CO_2 in the atmosphere.
Volcanism	*CO_2 is typically the second most abundant gas released during volcanic eruptions.*	*CO_2 is released from the lithosphere into the atmosphere and oceans.*	*An increase in CO_2 in the atmosphere enhances the greenhouse effect and warms the Earth's surface.*
Chemical weathering	CO_2 in rainwater combines with minerals in rock to form calcium carbonate.	Carbon flows from the atmosphere and oceans into the crust.	**Climate cools.** Carbon is being drawn out of surface environments and stored in the crust. Uplift of high plateaus and mountains may enhance this flux.
Metamorphism	Heating, recystallization, and decomposition of rocks during metamorphism can release large amounts of CO_2.	Carbon flows from the rocks (the crust) into the atmosphere and oceans.	**Climate warms.** Increased levels of CO_2 in the atmosphere enhances the greenhouse effect, which acts to trap heat energy and slow down the loss of heat to space.
Human activities: combustion of fossil fuels	The burning of fossil fuels releases large amounts of CO_2 into the atmosphere.	Carbon flows from the lithosphere (coal, oil, and gas) into the oceans and atmosphere.	*An increase in CO_2 in the atmosphere enhances the greenhouse effect and warm the Earth's surface.*

Answers to Review Questions

1. A. The Earth's atmosphere is about 75% nitrogen gas.
2. C. Acids are powerful agents of chemical weathering.
3. A. Carbon dioxide is transparent to visible light and absorbs heat (infrared radiation). Refer to Figure 23.6.
4. A. Refer to Figure 23.8.
5. B. Sulfur dioxide generated by burning fossil fuels high in sulfur content is the major source of acid in rain. The sulfur dioxide combines with rainfall which is already slightly acidic from the dissolved carbon dioxide to produce sulfuric acid.
6. A. Refer to the *Stratospheric Ozone Depletion* section in your textbook.
7. D. Refer to the *Stratospheric Ozone Depletion* section in your textbook.
8. B. Refer to the *Carbon Budget* section in your textbook.
9. B. Volcanic eruptions are not associated with El Niño events. Refer to *Earth Issues* box 23.1.
10. A. As the oceans become warmer, carbon dioxide solubility decreases. (Remember from Chapter 17 that carbon dioxide solubility increases as water temperature decreases.) Therefore, more carbon dioxide will be released, which is positive feedback. Positive feedback adds carbon dioxide to the atmosphere; negative feedback subtracts it.
11. C. Carbon dioxide is a greenhouse gas.
12. A. The greenhouse effect significantly influences surface temperatures for planets with atmospheres.
13. A. Latitudinal patterns in extinctions would be strong evidence that climate played a role in causing the extinction events.
14. C. Living organisms.
15. D. At present human fossil fuel burning contributes far more atmospheric carbon than does deforestation. The other alternatives (plant uptake and ocean gas exchange) represent in part a response to human activity. Normally there is a balanced carbon flow between plants and atmosphere and oceans and atmosphere. See Figure 23.11. Human activities are disrupting this balance, so the climate system responds by absorbing more carbon into the oceans and increasing plant production on land. Refer to Figure 23.12.